THE ENC[illegible]A OF THE
HARLEY DAVIDSON

HARLEY-
DAVIDSON

THE ENCYCLOPEDIA OF THE HARLEY DAVIDSON

An illustrated guide to an iconic motorcycle with 600 photographs

PETER HENSHAW & IAN KERR

HH
HERMES HOUSE

This edition published in 2010 by Hermes House, an imprint of Anness Publishing Ltd, Hermes House, 88–89 Blackfriars Road, London SE1 8HA, UK; tel. 020 7401 2077; fax 020 7633 9499

www.hermeshouse.com; www.annesspublishing.com

For all editorial enquiries, please contact Regency House Publishing at www.regencyhousepublishing.com

ETHICAL TRADING POLICY
Because of Anness Publishing's ongoing ecological investment programme, you, as our customer, can have the pleasure and reassurance of knowing that a tree is being cultivated on your behalf to naturally replace the materials used to make the book you are holding.
For further information about this scheme, go to www.annesspublishing.com/trees

CONTENTS

Who says you can't take a Harley off-road? An early Duo Glide takes to the water.

VR1000: Harley-Davidson's return to top-level racing wasn't a conspicuous success until the 1990s, but the work goes on.

THE HARLEY-DAVIDSON STORY

1900s: Early Days

In 1903, it must have seemed to Mr and Mrs Davidson that their three sons had gone their separate ways. None had followed their father into the carpentry trade, but all were holding down jobs in engineering. William, the eldest, had already done well for himself, well established as a foreman with the Milwaukee Road railroad company; Walter, the middle one, had moved away to Kansas to work as a machinist; and young Arthur remained in Milwaukee, earning his crust as a patternmaker at the Barth Manufacturing Company. Arthur, in particular, seemed to have a bright future ahead of him: he was an outgoing, personable young man with a lot of drive, not to mention a good friend named William Harley. The Davidson brothers' parents had emigrated to the USA from Scotland in around 1871, while Bill Harley's parents had come over from the North of England. All chose to settle in Milwaukee, Wisconsin.

It was hardly surprising that Bill and Arthur were soon turning their considerable energy towards the internal combustion engine. The world they grew up in was one of optimism and belief in fast-developing technology. Steam power had transformed Europe and North America, and now electricity and the small petrol engine promised to do the same.

Where it all began: the Harley-Davidson Motor Company's first 'factory'.

After centuries of reliance on horse or steam power, mankind was on the brink of discovering a new age in transport. Maybe it was no coincidence that, in 1903, the Wright brothers tested their first powered aircraft, the Model A Ford hit the roads, and the first Harley-Davidson motorcycle was sold to a paying customer.

Some say that the young Harley and Davidson wanted to build a small petrol engine to pace cycle races; others that it was to power a row-boat, both being keen cyclists and fishermen. (Incidentally, it was by mutal agreement that their company came to be known as 'Harley-Davidson' – Davidson-Harley just didn't sound right.) Whatever the true reason, from the winter of 1900 they began to spend much of their free time designing their first engine. They weren't on their own, though, as a workmate at Barth, a German named Emile Kruger, had practical experience of the De Dion engine, and Ole Evinrude, whose name was to adorn countless outboard motors in years to come, helped them to develop a suitable carburettor. That first 10-cubic inch (165-cubic centimetre) engine wasn't a masterpiece: speed was controlled by the spark setting – but it ran

A well known portrait of four very famous men. From left to right: William A. Davidson, Walter Davidson Snr, Arthur Davidson, and William S. Harley.

and it worked, so the two young men did the obvious thing and bolted it into a bicycle frame.

Meanwhile, Arthur had been writing to his elder brother, Walter, keeping him abreast of developments. Walter was evidently interested, for in 1902 he took a job in Milwaukee and moved back home, all the better to help in the exciting new enterprise. But 'enterprise' is perhaps the wrong word. There is no evidence that the three saw their prototype motor as the basis of a business empire; all kept their full-time jobs, and it wasn't until the spring of 1903 that the first prototype putt-putted its way onto the streets of Milwaukee. For many young entrepreneurs, this would have been the cue to rush into production: after all, it ran didn't it, after a fashion? Moreover, many early motorcycle makers were all too eager to jump straight into production with machines that worked, just; they didn't stay in business long.

Harley and the Davidsons were made of sterner stuff. Only when their prototype worked consistently did they consider selling a replica. In fact, this cautious, conservative attitude was to typify the way they did business over the next four

LEFT: Even in the early days Harleys were exported. This is an F-head outfit in India.

decades. So when the tiny prototype obviously needed more power, it was discarded and a much bigger 25ci (410cc) motor was designed in its place. This one had enough power, but soon vibrated the standard bicycle frame to pieces. The only answer was a purpose-built frame, with bigger brakes, wider wheels and heftier bearings. True, it still had pedals, which you had to work like crazy to start it, it cruised at around 25mph (40km/h) and had no gears or suspension, but the motorized bicycle was becoming a true motorcycle. Better still, unlike some of its more hastily-assembled rivals, the Harley-Davidson worked, and kept working. People started to take notice, and the first orders began to trickle in.

It was time to get serious, and while William Davidson Snr was building a 10 x 8ft (3 x 2.4m) shed in the back garden (the first Harley-Davidson factory), Bill Harley was leaving Barth to study for an engineering degree. Over the winter of 1903–4 Arthur built up two bikes for paying customers, while Walter left his job to concentrate on the new venture full-time. Once word got round that the Harley-Davidson was a reliable piece of machinery, more orders flowed in. A rich uncle, James McLay, lent them enough money to build a proper factory on Chestnut Street (later renamed Juneau Avenue), where in 1906 49 bikes were produced. Harley-Davidson motorcycles emerged from that site for the next 70-odd years, and even now the famous V-twin motor is still made there. Production grew in leaps and bounds as the orders kept coming, and more staff was taken on to cope: output tripled to 152 bikes in 1907, then tripled again the following year. By now, Arthur, who was a born salesman, had left his day job and was out on the road demonstrating the bike and recruiting dealers, and his eldest brother, William, came on board as works manager. Harley-Davidson Incorporated was born.

OPPOSITE: Spot the difference? That's a 1905 single on the floor and a 1907 version on the bench.

HARLEY-DAVIDSON
HARLEY-DAVIDSON

LEFT: A 1905 F-head single, when bikes were still bikes. The pillion seat is non-standard and by the look of it, only for the brave!

1910s: Rapid Expansion

Meanwhile, the motorcycle had hardly been changed since 1903. It was now nicknamed the Silent Gray Fellow (it was quiet, and came in only one colour) but was much improved by the addition of sprung leading-link forks, which Bill Harley designed while at college. Incidentally, the same basic design of forks was used by the company until 1947, and was even reintroduced (in modernized form) in the late 1980s. Soon after, the 25ci (410cc) engine was enlarged to 35ci (575cc), housed in a larger, longer wheelbase frame. Walter proved the latest version by winning the Long Island Endurance Run in 1908, and consolidating it with an economy run win the week after.

But though the 5-35, as it was officially known, could top 50mph (80km/h), it still wasn't enough for the emerging motorcycle riders and the huge distances they travelled in America. Like all its rivals, Harley-Davidson needed a bigger bike.

Strange as it may now seem, Bill Harley did not invent the V-twin: when he paired two existing singles onto a beefed-up crankcase, he was merely doing the same as everyone around him. The V-twin was a relatively quick and easy way to build a more powerful engine using existing parts, though Harley's first prototype seemed hardly that. Still hampered by an atmospheric inlet valve (opened by piston suction, rather than positively via a camshaft and pushrod), it was no faster than the single – the 49-ci (803-cc) twin was quickly withdrawn from sale, and Bill went back to the drawing board.

When it reappeared as the F-type in 1911, it was clear that he'd been working

Harley-Davidson's image was somewhat different in the early days, as illustrated by this solid and reliable sidecar tug.

hard. It now had a mechanical inlet valve, so it could rev higher and produce more power, and the drive belt was tensioned to prevent slippage (another problem with the prototype). In some ways, the F-type, with its fixed belt drive and gravity-fed lubrication, seemed little advanced on the early pioneer motorcycles, but this was about to change. The years 1910 to 1915/16 saw a brief flurry of innovation from the American motorcycle makers which, for a while, put them ahead of the Europeans: and Harley-Davidson was right in the thick of it. A basic clutch in the rear hub allowed the rider to stop and restart without stopping and restarting the motor. That was soon overtaken by a proper multi-plate clutch and chain drive with a two-speed gear in the hub. This was a real breakthrough, which Harley followed up in 1915 with a three-speed gearbox and automatic engine oil pump. Also that year (they must have been working nights) was the option of electric lighting and fitted with this, the F-model became the J, using a gear-driven magneto-generator.

Harley-Davidson's success mirrored these technical leaps forward, and production continued to soar: over 3,000 bikes were sold in 1910, over 5,000 the year after and 9,000 the year after that, while over 16,000 left the factory the year war broke out in Europe and 22,000 as the decade came to a close. Only ten years after that first prototype first hit the road, Harley-Davidson had established itself as number two in the American motorcycle market; the longer-established Indian still led, with Excelsior a poor third. In fact, it was the start of an intense, sometimes bitter, rivalry between Indian and Harley-Davidson, which lasted for the next 25 years. Harley-Davidson certainly did better out of the First World War than its rival. In a fit of patriotic fervour, Indian turned over its entire production to military needs with the result that there were many disgruntled Indian dealers with no bikes to sell; Arthur Davidson lost no time in persuading them to change sides.

This Indian/Harley rivalry was never more obvious than at the racetracks which included the wooden boardtracks of the early years and the later dirt ovals. Harley and the Davidsons weren't natural race goers, but they were shrewd enough to realize that success on the racetracks could translate into sales, and engineer William Ottaway (a man central to Harley-Davidson's engineering policy for many years) developed the short wheelbase II-K racer out of the J-model. It could top 90mph (145km/h), and win races, so in 1916 Ottaway got the go-ahead for the Model 17 V-twin racer, with four valves

OPPOSITE: A glimmer of progress? Still with belt-drive, but sprung forks have arrived.

LEFT: A 1909 single, much as it would have left the factory.

BELOW: 1909 and the first twin, which was hardly a success.

per cylinder. With the help of cylinder-head specialist Harry Ricardo, the 8-valve produced 55hp and in the hands of the official factory team (the delightfully named 'Wrecking Crew') it was soon notching up victories all over America. However, Harley's official racing effort was short-lived as the collapse of the motorcycle market in 1920 caused the whole team to be disbanded.

Most of those racers were V-twins (though there were some four-valve singles), but that simply reflected the American motorcycle market. Harley's own V-twin had long since overtaken the little single in importance, which was drifting into the background. The availability of cheap cars (not the least of which was Henry Ford's Model T) meant that motorcycles were increasingly seen as leisure tools for sporting young men, and a few women, and they wanted power and speed. However, various motorcycle makers still believed that a quiet, reliable and cheap machine could persuade the man in the street back onto motorcycles. Harley-Davidson's contribution was the Sport Twin of 1919. The 'Sport' part of it was a misnomer, as the new bike was a docile, mild-mannered flat-twin which could barely top 50mph (80km/h). On the other hand, it was quiet and easy to ride, had an enclosed chain for practicality, a modern three-speed gearbox and the option of electric lights. Unfortunately, it failed to attract the masses back to motorcycling and was dropped after a few years. It was quite well received in Europe, owing much to the English Douglas in its design, but can be regarded as Harley-Davidson's first flop.

This was bad news, as the Sport Twin was part of a massive expansion plan based on borrowed money. Since 1903, Harley-Davidson sales had spiralled upwards year after year, to the point where it was catching Indian, the market leader. So in 1919–20, the four Founders appeared to forget their natural caution and borrowed $3 million (a huge sum at the time) to expand the factory. When it was finished, the Juneau Avenue works was the largest motorcycle plant in the world, and produced a record 28,000 machines in 1920. Now in prosperous middle age, it seemed that Arthur, Bill, Walter and William could do nothing wrong.

ABOVE LEFT: A 1913 twin, now with chain drive.

ABOVE: A 1913 single: pedals were still there to start it.

LEFT: A racer from 1914, Harley-Davidson's first official year on the track.

RIGHT: Silent Gray Fellow, quiet and lacking in colour like a retired bank clerk.

ABOVE LEFT: An unrestored 1913 V-twin.

ABOVE: A single from the same year.

LEFT: A 1914 twin with pillion seat and painted rims.

RIGHT: Pushrod-operated inlet valves were the key to the early twin's success.

ABOVE, LEFT and RIGHT: Most early Harleys were destined to haul sidecars, but some were sporting solos.

OPPOSITE LEFT: Typical 1915 riding gear: mackintosh, overtrousers and flat cap.

OPPOSITE RIGHT: Boardtrack racers were stripped to the bone. This one dates from the 1920s.

1920s: Down & Up

Although the four Founders had become accustomed to success, they had to cope with economic failure in double-quick time. Harley sales more than halved in 1921 (to 10,202), which left them with an inventory of unsold bikes and a $3 million loan to pay off. But to to have achieved the success they already had, the four had needed plenty of business acumen. They shut the newly expanded factory for a month, cut salaries by 15 per cent and abandoned the racing programme. So sudden was this decision that the racers were actually left stranded at the State Fair races at Phoenix. They had to pawn their wristwatches and borrow money from a local Harley-Davidson dealer in order to buy their train tickets home!

This abrupt belt-tightening worked: the unsold bikes found buyers and production began to recover. But it was a fluctuating recovery; the motorcycle boom time was over, thanks to cheap, affordable cars, and it took the company 20 years to match that peak of 1920. What the new generation of riders wanted was more power, and Harley-Davidson obliged with the Superpowered Twin of 1922. Although intended as a sidecar machine, the extra power from its enlarged 74-ci (1200-cc) V-twin was welcomed by keen solo riders. The engine was no more than a bigger version of the original twin, still with overhead inlet valve, side exhaust and total-loss lubrication. Like its predecessor, the bike came in magneto (FD) or electric light (JD) forms.

Harley-Davidson had always set a great deal of store by its dealers, taking time to build up a network when less patient rivals were selling bikes to anyone with cash. Now, as one of the two makes dominating the market, Harley-Davidson was able to tighten its control over dealers: factory representatives toured the country, checking up; franchise agreements were for one year only (screw up, lose your franchise); and no dealer was permitted to sell any other make alongside Harley-Davidson, even the little Cleveland, which

sold to a completely different market. Of course, if you toed the line, there were rewards to be had, as the Harley-Davidson had a good reputation and sold well; but if you didn't, you wouldn't be a Harley-Davidson dealer for very much longer.

One thing that restricted dealers' freedom was that Indian and Harley were the only two manufacturers selling bikes in large numbers. And despite the fierce rivalry between them, the two actually met regularly to fix prices. It was of course illegal, and had to be done in secret, but in 1922 Frank Weschler (of Indian) and Arthur Davidson met and agreed to sell at the same price the following year; it became an annual event in their respective diaries. But Harley-Davidson was really in better shape than Indian by the early 1920s, whether or not they were selling more bikes. It was more efficient, had a tighter grip on its dealers and a simpler model range.

Nevertheless, Indian often seemed ahead on model development: it had been first with a flat-twin (which Harley-Davidson countered with the Sport Twin) and in 1925 launched the 21ci (350cc) Prince, a single-cylinder machine in the British mould. A Prince, a 350cc BSA and a New Imperial were shipped into Milwaukee for evaluation, and soon after Harley-Davidson's own 350cc single appeared in 'A' (side-valve) and 'AA' (overhead valve – ohv) guise. The former sold well as a basic utility bike, while the AA (nicknamed 'Peashooter') did very well in racing. It also highlighted another aspect of the Harley/Indian wars: Indian was invariably first in bringing out a new model, which Harley-Davidson would soon counter. Not only the Sport Twin and

THIS PAGE: The 1922 Sport Twin was a real departure for Harley-Davidson but, apart from export sales, was not a success.

OPPOSITE: This 1922 sidecar outfit is in rare original unrestored condition.

Peashooter, but all of Harley's new bikes of the 1920s – the Two Cam, 45 and VL – were all direct responses to what Indian was already making.

The Two Cam became something of a legend. The problem for Harley riders was that Indian side-valve twins were far faster than the older ioe (inlet-over-exhaust) Harleys, even if they didn't have the same stamina. The Two Cam answered this by using a separate cam for each valve, thus allowing tuning for more light, higher compression, higher revs and more power. Available in 61ci (1000cc) and 74ci (1200cc) forms, it wasn't cheap ($50 more than the equivalent J) but it was said that a good 74 would reach 100mph (161km/h). No longer need Harley-Davidson riders be embarrassed on club runs! In fact, the Milwaukee machines, despite their earlier racing success, had gained a reputation as old men's machines – reliable and dependable, but without the performance of an Indian or Excelsior. The 45ci (750cc) Indian Scout was a particular problem for Harley dealers, and the Founders' response (which Walter Davidson had promised to shareholders in 1927) was the model D, better known as the 45. Like the Scout, it was a 45ci side-valve V-twin. Unlike the Scout, it was slow and gutless, unable to reach 60mph (97km/h) where the Scout could top 75. A carburettor kit had to be rushed into production to improve power, and to add salt to the wound, the distinctive vertically-mounted generator was prone to failure. The 45 had clearly been rushed into production without proper testing. In time, it developed into a

1922
HARLEY-
DAVIDSON
H-D
DUNLOP

thoroughly reliable machine, powering the wartime WLA and the three-wheel Servicar. But it was a long way from the care and diligence with which the first single had been developed. Was Harley-Davidson losing its way?

It must have seemed so to loyal riders and dealers when the new VL was unveiled in 1929. This replacement for the long-lived F/J models was almost all-new, the bore, stroke and capacity being all that remained the same on this side-valve V-twin for which the factory claimed a 15 to 20 per cent power improvement. Unfortunately, the real increase was just one horsepower, and to make matters worse the VL weighed a whole 120lb (54kg) more than the F/J. In a bid to overcome this lethargic performance, the engine had been given relatively small flywheels which did allow good acceleration up to 50mph (80km/h) or so, but that was about it. Top speed and hill-climbing ability were sorely lacking on this big twin, and the light flywheels also caused finger-numbing vibration. Some customers wanted the F/J and Two Cam reinstated; others wanted their money back.

As with the 45, Milwaukee had to come up with answers, and fast. After much burning of the midnight oil, the solution came in bigger, heavier flywheels and modified cams. This did the trick, but the bigger flywheels needed bigger crankcases which in turn demanded a larger frame. In other words, every VL made up to that point (there were 1,300 of them) had to be completely rebuilt. It cost Harley-Davidson $100,000 in new parts, caused much heartache to the dealers (who were expected to do the rebuilding at their own expense), and did much damage to the hard-won reputation for reliability. Better was to come, but there were more hard times to get through first.

THIS PAGE: It's 1925 and the J-model with overhead inlet, side exhaust valves has four years to run. It had electric lighting and a three-speed hand-change gearbox.

OPPOSITE: This J sidecar outfit in its original green was undeniably handsome. The substantial luggage rack was standard but the front brake was not.

Harley-Davidson

1930s: Slump to Recovery

Twice in its lifetime, Harley-Davidson has come close to closure. The most recent was 1985 when its major financial lender pulled out. But 50 years before, there had been talk in the boardroom of voluntarily throwing in the towel. With the benefit of hindsight, this seems odd: the company was challenging Indian for market leadership and had overcome the sales slump of 1920 and some troublesome new bikes, which were now settling down nicely. The reason was more fundamental than any of this. The slump following the Wall Street Crash of 1929 had hit motorcycle sales hard: Harley sold nearly 21,000 bikes that year, over 17,000 in 1930 but a mere 3,700-odd in 1933. It was a disaster. The four Founders were now old men; they and their families were financially secure and they'd worked hard all their lives. Why carry on? Well they did: whether it was loyalty to the workforce, some of whom had been there since the start, or simply a refusal to throw away 30 years of work. Bill Harley and the Davidsons decided to stick with it. Not only that, but they put a lot of money and time into an all-new bike that for them was a real leap into the unknown. This was the Knucklehead, a real milestone bike for the company that finally enabled it to leap ahead of Indian.

But in the meantime, men were laid off, salaries were cut by 10 per cent and the Founders reduced their own remuneration by half. The side-valve VL and 45 were gradually improved, the 45 in 1932 with a redesigned frame that did

ABOVE LEFT: By the late 1920s, Harley-Davidson could claim that its Juneau Avenue factory was the biggest in the world.

ABOVE: The 1937 National Hillclimb Championship in which Harold Seamans, competing in the 80-ci (1300-cc) Class B event, breaks the winning tape while riding a factory bike.

LEFT and BELOW: 1928 was the last year for the J model, the engine shot clearly showing the pushrod inlet valve and side exhaust valves. This motor served Harley for nearly 20 years and not everybody approved of its replacement.

away with that odd vertical generator, which had led to the derogatory phrase, 'three-cylinder Harley'. Renamed the R-series, it was also given a boost when Indian temporarily dropped the Scout which had been taken over by the du Pont concern, in a move towards rationalization. The upshot was that the 45 had the middleweight market to itself. Meanwhile, the genius of tuner Tom Sifton and others was coaxing it into winning a few races, to the extent that the factory later produced its own 45 racer, the WR.

Another sign of faith in the future was the introduction of a range of bright new colours: for years, Harleys had only come in dark, drab hues that reflected their workaday image. Now, there were eye-catching reds, oranges and creams, two-tones and stylish art deco tank logos. Once again, Indian was the spur to action. The du Pont empire was based on chemicals, which gave Indian access to a whole range of paints. Harley-Davidson had no choice but to follow suit. The same went for the VLH of 1935. Indian had a big-engined 80ci (1300cc) V-twin, the Big Chief, which Harley-Davidson did not. This would never do, and the VLH had its crankpin moved outboard to allow a $4\frac{1}{4}$-in (108-mm) stroke and the biggest Harley motor yet.

But while all this was going on, the engineering department (such as it was, manpower having been depleted by the various cuts) was hard at work developing the Knucklehead. It actually took five years to reach production, partly because of the skeletal engineering department,

partly because of teething troubles with oil leaks and the new dry sump lubrication system. When it was finally unveiled in 1936, at the annual dealer conference, everyone seemed to know it would be a hit. It was new from stem to stern: the 61ci (1000cc) overhead-valve V-twin had a shorter stroke than its predecessors, so it was able to rev higher and harder than any previous Harley. It had a proper recirculating oil system in place of the outmoded total-loss system, which meant a constant supply of fresh, cool oil, better able to cope with high-speed runs. It had big valves and a high compression. It had 40bhp at 4,800rpm and could top 90mph (145km/h). There was a new four-speed constant-mesh gearbox, new clutch, new frame, new swoopy styling. The '61E' was its official title, but it was nicknamed 'Knucklehead', suggested by the shape of the rocker boxes.

Even so, the Founders, now increasingly assisted by their respective sons, tried to insist that the new bike was for limited production only, and dealers were forbidden to order them as demonstrators. Maybe they were wise, for the Knucklehead didn't have a trouble-free first year: it went on leaking oil until valve spring covers were introduced, and the springs themselves broke, but these really were teething problems. In any case, Joe Petrali set a new speed record with one of the new bikes of 136.15mph (219km/h), thanks to a tuned engine and high top gear, while police officer Fred Hamm broke the 24-hour record, riding a standard Knucklehead 1,825 miles (2937km) in 24 hours (an average of just over 76mph/122km/h, including stops). It was official, Harley-Davidson finally had a bike that was both reliable and fast.

The Knucklehead's success rubbed off on the other bikes. In 1937 the 45, 74 and

Art deco comes to Milwaukee on this 1932 VL, but Harley had Indian to thank for the new range of bright colours. The new sidecar motor proved a reliable performer after a troublesome start.

80ci (750, 1200 and 1300cc) side-valve all received the new lubrication system (which was just as well, the total-loss being a museum piece by now), and the new bike's streamlined styling. The 45 became the W-series, and the V was now the U-series.

While the Knucklehead stole the headlines, something completely different was doing good business for the company. The three-wheeled Servicar used the front end of a 45 married to a two-wheel rear end and a load-carrying box. Originally intended for garage mechanics to collect cars for repair, it proved ideal for traffic police and meter maids, becoming something of an American institution. It remained in production until 1973.

ABOVE LEFT and RIGHT: The neat 1934 '45' was more often known by the number than the official 'D'- or 'W'-series. Never fast or glamorous, it was nevertheless a tough and reliable machine that introduced thousands to motorcycling in the 1930s and 1940s.

LEFT: A late-1930s VL, ready to tour. Note the large buddy seat, panniers, buckhorn bars and luggage rack.

HARLEY-DAVIDSON

OPPOSITE: A rare VL of the early 1930s. Mirrors and buddy seat were extras.

TOP LEFT: A 1934 VL.

TOP RIGHT: The 1935 VLU was at 80ci (1300cc) the biggest Harley yet.

LEFT and ABOVE: Mid-1930s Servicars with optional large carrying box.

A study in changing attitudes: in the 1920s Harley-Davidson motorcycling was portrayed as a peaceful, almost pastoral activity. Ten years later ('The Greatest Sport of Them All'), the image is altogether more dynamic while by the 1940s power and performance had become key words, with riders' apparel taking on a distinctly militaristic flavour.

ABOVE: Four jackets from *The Enthusiast* magazine, featuring touring (1939), racing at Daytona Beach (1940), a military machine (1942) and an optimistic look towards peacetime (1944).

LEFT: A charming simple image from the 1930s.

1940s: Wartime Diversions

Just a handful of years after the four Founders contemplated closing down Harley-Davidson, the family firm was back on the crest of a wave. It had survived two major slumps and the difficult, competitive years between them, and had finally overtaken its arch-rival and transformed itself from producer of plodding side-valves into a maker of fast performance machines. The Second World War was to keep the factory busier than ever, with thousands of bikes and countless pieces of subcontract work.

But in 1940, the USA hadn't yet declared war on Germany, and plans for next year's peacetime Harleys could go ahead. The 80ci (1300cc) side-valve received aluminium cylinder heads in 1940, or better cooling, and 1941 saw a 74ci (1200cc) version of the Knucklehead, whose new centrifugally controlled oil pump finally solved the old problem of too much lubrication in some places, not enough in others. Not many of them reached civilian hands though. Also under discussion was a 45ci (750cc) overhead-valve engine, which was logical, given the success of the Knucklehead. It had been discussed for some years, and Bill Harley was a strong advocate of the idea. Three prototypes were built, one of which was taken on a 5,000-mile (8050-km) proving ride into Texas and back. There were no major problems, and the new baby Knucklehead looked promising; but it would have cost as much to build as the bigger side-valves, and the idea was dropped. It would be nearly 20 years before Harley-Davidson finally produced an overhead-valve 'baby' V-twin – the Sportster.

In any case, even by the late 1930s, the Board had had more pressing things to consider, notably the potential of big orders from the army for a standard do-everything motorcycle. Harley's solution was the WLA, basically a militarized version of the standard W-series 45. With its low compression motor, big oil bath air filter and off-road skid plate, the WLA was a truly rugged workhorse. The extra weight of all the army equipment depressed top speed to around 50mph (80km/h), but that hardly mattered unless you were being chased by BMW-mounted Germans. What did matter was that nearly 90,000 were supplied to the forces, along with a WLC version for the Canadians, which introduced thousands of servicemen to motorcycles in general and Harleys in particular.

There were plenty of other military projects at Milwaukee, not all of which saw the light of day. The XA *was* produced, basically a copy of the contemporary BMW, because that's what the army wanted. A civilian prototype of this flat-twin, shaft-drive bike was actually running in 1946, though it came to nothing. The XA motor was also mooted for a mini-Jeep, which could be parachuted

OPPOSITE: The 1942 XA was a strictly military machine. Few were made.

LEFT, BELOW LEFT and BELOW: The Knucklehead heralded a new era of style for Harley-Davidson, quite apart from being the fastest, most powerful Harley for the road yet seen. These examples all date from 1938/39 and are of 61ci (1000cc) capacity.

into battle zones, for a generator set, and as an updated modernized Servicar.

While all this was going on, a new generation of Harleys and Davidsons was taking over. William Davidson had died just before the war in Europe began; Walter Davidson died in February 1942 and Bill Harley the year after. Arthur, the sole survivor of the four Founders, was killed in a car crash in 1946. But William's son (another William) took over the helm of the company, while Harley's son controlled the engineering side: Harley-Davidson was still very much a family firm as it faced a very different post-war world.

For the first time it was to encounter a significant threat from imports. Harley-Davidson now accounted for 75 per cent of US motorcycle production, but there was a growing trickle (soon to become a flood) of British motorcycles being shipped across the Atlantic. Indian tried to meet these head-on and failed. Harley-Davidson preferred to pretend that the threat did not exist. So instead of coming up with a mid-range competitor for the Brits (a modern vertical twin, with foot gearchange and rear suspension, was discussed), the company concentrated on what it did best, big V-twins. For 1947 the Knucklehead became the Panhead (the rocker covers now looked like upturned saucepans) with aluminium cylinder heads for cooler running and hydraulic tappets for quietness. In a now-familiar scenario, the tappets proved troublesome at first, with

ABOVE LEFT: Some Knuckleheads did reach the civilian market in the 1940s.

ABOVE: A highly though possibly over-restored 1942 WL45 and 1958 Duo Glide.

fluctuating oil pressure upsetting the valve timing; but overall the Panhead was a real improvement. There was even bigger news the following year when Bill Harley's 1907 Springer forks were dropped in favour of conventional telescopic ones, and the bike was named Hydra-Glide in their honour. The first of the 'Glides' had arrived. Gradual updates followed, with foot shifting (1952), rear suspension (1958), 12-volt electrics (1964) and electric start (1965).

ABOVE RIGHT: A 1940s Knucklehead. It was necessary to go easy on the chrome as wartime shortages made themselves felt.

RIGHT: 1948 marked the first year of the Panhead, with hydraulic tappets and alloy cylinder heads.

So it was business as usual, except for the S-125, or Hummer, which went on sale in 1947. This little 7-ci (125-cc) two-stroke was basic in the extreme, with 3bhp, a three-speed gearbox and suspension worked by rubber bands. It was really a pre-war German DKW, offered to the US as war reparations. Indeed, BSA got it as well, as the Bantam, and there was even a Russian version. In its Harley-Davidson guise the ex-DKW was actually quite a success, at least at first. Although some traditionalist Harley dealers refused to sell what they regarded as a pipsqueak machine that didn't deserve to be called a motorcycle, it sold in steady, though not spectacular, numbers for over a decade, by which time it had grown to 10ci (165cc) and had telescopic forks.

OPPOSITE: A springer fork marks this out as an early (1948) Panhead.

LEFT: The 1946/47 Knucklehead was hardly changed from its pre-war version.

BELOW LEFT and RIGHT: By 1946 the Knucklehead had become the mainstay of the range.

1950s: A Challenge from Imports

The Hummer was all very well, but it wasn't the middleweight Harley for which dealers were clamouring. Faced with a nimble 30-ci (500-cc) Triumph or Norton, the ancient 45 looked like an overweight dinosaur. Throughout the 1950s, Harley held on to its traditional club and touring market, selling about 5–6,000 big twins a year to these riders. But it was missing out on the new sports bike market; by 1950, imports had captured 40 per cent of US sales, prompting Harley-Davidson to ask for a swingeing import tariff to be imposed. At the subsequent hearing, some unpleasant facts about how the company had restricted its dealers' freedom of action came to light and the request was refused.

Harley's answer to the imports was unveiled in 1952, and the new K-series, which finally replaced the long-serving 45, was a real enigma. Why? Well on one hand, it had an up-to-the-minute chassis, just like the imports it was designed to compete against: telescopic forks, swinging-arm rear suspension, unit-construction gearbox with foot-shift, it was all there. But its power unit was a development of the old side-valve 45, though opinion is divided among Harley historians as to how new it was; the K-series really did look like old wine in a new bottle. Side-valves were outmoded before the war, so it was mystifying that Harley-Davidson stuck with them into the 1950s. It wasn't as though the firm was short of expertise (its prototype ohv 45s had been running in 1940); nor was it short of money (it spent $3.5 million on new plant in the late 1940s); it wasn't short of time to develop a new engine; and finally, there really was an all-new prototype to meet the imports, with an all-aluminium 60-degree V-twin with ohv, twin carburettors and high cams. Whatever the reason, Harley dealers were stuck with the K-series, whose sales matched their disappointment. It could barely exceed 80mph (129km/h) where the Triumph Thunderbird could top 100. Things were improved after a couple of years when Harley boosted the capacity to 55ci (883cc), but it was still no Thunderbird beater.

Meanwhile, the whole of motorcycling in America was changing, and not just because of British imports. As we've seen, Harley's traditional club market carried steadily on, which was an organized, orderly sport, with clean-cut men and women wearing smart uniforms. But there was a new strand of biking on the way which had first come to public attention back in 1947, thanks to the supposed 'riot' at Hollister, California. The new bike gangs that were forming at the time, often composed of ex-servicemen, had a very different ethos to the orderly clubs. For them, motorcycling was at the centre of an outlaw lifestyle which rejected mainstream

OPPOSITE and ABOVE: Telescopic forks and hardtail rear end add up to the Hydra-Glide Panhead.

RIGHT: The oddball 1955 KHK had a modern chassis with a side-valve motor.

society. Their bikes reflected that. Instead of the traditional Harley 'dresser', with panniers, screen and crash-guards, the bike of a member of Satan's Sinners had every piece of superfluous equipment stripped off to cut weight and boost performance. But in time, it became a statement in itself. This was the start of the custom bike movement, and Harley-Davidson, much as it tried to distance itself from the outlaw clubs, was right at the centre of it. Ironically, it has since embraced a highly

LEFT: Harley-Davidson is keen on birthdays. This is a 50th-Anniversary Hummer from 1954.

OPPOSITE: The Hummer eventually spawned an off-road version, the forerunner of the trail bike.

stylized, sanitized version of the same thing. The difference is that new Harleys are sometimes bought by 'weekend outlaws', who wear black leather at weekends but are back in suit and tie come Monday morning.

The tourers and custom bikes are two distinct families of bikes within Harley-Davidson's current range. The third is the Sportster, which made its debut in 1957. Just like the Knucklehead of 1936, this was a milestone bike for Milwaukee. There was actually nothing radical about it, based as it was on the running gear and basic engine unit of the K-series. What it did have was a new top end with overhead valves and a 7.5:1 compression, which produced 40bhp at 5,500rpm. It wasn't enough to beat the 40ci (650cc) British twins, but it did equal them. That was taken care of the following year with the higher-compression XLH Sportster that claimed 55bhp at 6,800rpm. Once again, Harley finally had a bike that was as fast as anything else on two wheels.

The other thing it gained that year was a look we all associate with the Sportster. Its origins lay in the stripped-down XLC (no lights or battery) of which just 200 were built for competition. The road-going XLCH aped that style, with high-level staggered pipes and a tiny 2.2-gallon (10- litre) fuel tank taken from the Hummer 125. It may not have been as practical as the standard Sportster, but it looked like a lean all-engine machine, and it outsold the frumpy XL by two to one. But even with the XLCH, the Sportster wasn't an instant hit. It was actually outsold by the big twins until 1969, and throughout the early 1960s, less than one in five Harley sales was a Sportster. However, in the long run, that didn't matter, as the Sportster opened up a new market for Harley-Davidson, one which is still going strong.

423·T
MC-OHIO-51
THUNDER HUMMERS

1960s: Takeover

The Sportster had finally met the British bikes head-on, at least in straight-line speed, but there was another new threat on the horizon. Honda and Suzuki had started to export to the US, and these clean, sophisticated lightweights looked as if they would open up another market which was closed to the company. It did have the 10-ci (165-cc) Hummer, but that was now over ten years old as well as based on a pre-war design; so to meet the Japanese something else was required. It is hard to credit now, but right through the 1960s Harley-Davidson sold lots of small motorcycles, even a 3ci (50cc) moped and a scooter! These came from two sources: developments and updates of the original Hummer, and bikes built by Aermacchi in Italy.

ABOVE and OPPOSITE LEFT: The 1959 FLH Duo Glide, by then the definitive American touring motorcycle. Screen and fibreglass panniers are typical period extras.

OPPOSITE RIGHT: Balloon tyres, soft long-travel suspension – riding doesn't get much more comfortable than this.

The Italian connection was made in 1961. Two years earlier, Harley-Davidson's own development committee had recommended that the best way to market a lightweight motorcycle was to import someone else's machine and stick the Harley-Davidson badge on it. So that's what they did. Aermacchi had been making aircraft during the war, producing scooters after 1945, then a range of four-stroke singles from the mid-1950s. An 80mph (129km/h) 250, with its distinctive horizontal cylinder, fine handling and light weight was on offer by 1957 – just the thing to offer Americans who hopefully wanted to buy a miniature Harley. Instead of just importing the bikes, Harley-Davidson decided to buy Aermacchi lock, stock and barrel, and in 1961 began to import the 250 as the Harley-Davidson Sprint. It had skimpier mudguards and a smaller fuel tank for the US market, and *Cycle World* was quite impressed when it tested one. The pushrod single would happily rev to 7,500rpm, it reported, and cruised at 65–70mph (105–113km/h), though even then it was able to assert that there was nothing very high-tech about it. Sure enough, the Sprint was technically overtaken by Japanese bikes a few years later, but it nonetheless made a valuable contribution to Harley-Davidson's market share.

Not that it came without headaches. Although mechanically reliable, the Sprint was not as well made as home-grown Harleys, and the electrics in particular gave a lot of trouble. It was bad enough that

Harley-Davidson engineers and managers were trying to co-ordinate with a factory thousands of miles away that worked in a different language, but Aermacchi also bought in many of its components from other factories. From Milwaukee, getting on top of Aermacchi quality proved an almost impossible task.

In any case, head office had its own plans for the lightweight market, which slotted in under the 15ci (250cc) Sprint. First came its assault on the scooter market. The Topper, which appeared in 1960, used the Hummer's 10-ci (165-cc) two-stroke engine in restricted (5hp) or unrestricted (9hp) guise, allied with automatic transmission and full bodywork. It was not an elegant machine, lacking the perkiness of a Vespa or Lambretta, nor was the Topper cheap at $600. And despite encouraging sales in its first year (when 3,800 were sold), and a cameo role in a popular TV series, the Harley-Davidson scooter was not a success, and was dropped in 1964.

A similar fate awaited Harley-Davidson's updates on the Hummer theme; with a slightly larger 175cc engine, it became the Pacer in road form, or the Scat with off-road-style high-level exhaust and knobbly tyres. There was also the Ranger, a true off-road bike with no lights or front mudguard, higher ground clearance and lower gearing. But all three of these bikes shared that elderly two-stroke engine, and a basic frame without rear suspension. When the Japanese lightweights arrived, they underlined just how crude the Harley tiddlers were and Milwaukee dropped them in 1966.

But if Harley's home-grown lightweights were failing to set the world on fire, the Sportster was a different matter. In small towns across the continent, the favourite Saturday night pastime consisted of an impromptu drag race between Harley and Triumph/BSA/Norton. A horsepower race ensued: Triumph produced the twin carburettor Bonneville, and Norton the 45-ci (750-cc) Atlas. So Harley-Davidson upped the 'touring' XLH Sportster to 55bhp, which in standard form could run a 15.5-second quarter-mile and almost reach 100mph (161km/h). It still looked like a mini-Duo Glide, but better was to come.

By 1968, with yet more power (now up to 58bhp), new front forks and a 114-mph (183-km/h) top speed, the XLH Sportster was living up to its name; even *Cycle World* was full of praise for its handling, which three years previously it had likened to an automobile's. Also, it at last acquired the classic Sportster tank. Who cared if you needed to fill up every 80 miles?

It was at a time when the Sportster was a serious rival to the imported big bikes, particularly in the later 1960s, when its weight disadvantage disappeared beside new heavy superbikes such as the BSA Triumph triples and Honda 750 four. It wouldn't last, of course, as the performance race soon left the Sportster behind in the early 1970s, but by then the 'baby Harley' had acquired a following all of its own, just as loyal as the big twin aficionados, and for whom ultimate performance was less important than the way the Sportster looked and felt.

A baby Harley of a very different sort was the M50 moped, another import from Aermacchi which was sold in America. By the standards of the time, it was a reasonable machine, with rear suspension and an attractive price tag. And the buyers seemed to agree. It seems a sacrilege now, but in 1965 and 1966 Harley's best-selling machine was a moped. The success was short-lived, however. After a couple of years the M50 was treated to a boost to 4ci (65cc) and a big advertising campaign, yet the 65cc

ABOVE: The Aermacchi Sprint: its high-level pipe, bashplate and high-rise front mudguard indicates an off-road specification.

OPPOSITE: The Electra Glide motor, shown here in early Panhead form.

Harley never achieved the popularity of the M50.

The Sprint, on the other hand, just seemed to sell better each year, despite its relatively low-tech approach. For one thing, Aermacchi/Harley-Davidson seemed able to come up with more power and new variations on the theme year after year.

The 1966 CRS was a serious scrambles version, with 28bhp at 8,500rpm (it could rev safely to 9,500) and no lights or speedometer. In racing CR-TT form it made 35bhp at 10,000rpm and a streamlined Sprint set up a new 250cc record, at 176mph (283km/h) on the Bonneville Salt Flats.

That record, set in 1965, helped to boost Sprint sales to 3,000 a year and they rose to 5,000 in 1966 and 9,000 the year after. A 21-ci (350-cc) version appeared in 1969, though in a less frenetic state of tune, with 25bhp. Most of the Japanese opposition was faster, but the Sprint was becoming something of a niche bike, offering a different riding experience. It certainly went on selling a steady 4–4,500 a year right up to 1974. For a basic pushrod single with a four-speed gearbox and roots going back to the mid-1950s, it had a remarkable performance.

With all these small bikes coming and going, the big twins were rather left to fend for themselves in the early 1960s, but change did come mid-decade. Twelve-volt electrics in 1964 prefaced the fitting of electric start the following year, coupled with one of motorcycling's all-time classic model names – Electra Glide. Although regarded as a new model in its own right, the Electra was really just the standard FLH with push-button starting, though the name at least has proved very durable. It also added 75lb (34kg) to the weight, and to cope, the engine was updated into the Shovelhead in 1966.

As before, this was evolution, not revolution, in this case, the fitting of a Sportster-style top end to the big twin, which gave a useful power boost to 60bhp. It went up again in 1968, to 65bhp at 5,400rpm, thanks to a Tillotson carburettor and improved porting. Otherwise, it performed in much the same way as any Harley big twin before or since, happy to rumble along at 70mph (113km/h) until the tank ran dry. At the time, the Japanese had yet to come up with their own V-twin Harley lookalikes, so the Electra Glide was still offering a unique riding experience.

This was fine for the traditionalists, but Harley-Davidson was getting further and further out of the main stream. So while the motorcycle market expanded around it, the company was getting pushed into a backwater. The Aermacchi bikes had increased sales and turnover, but not profits, and the company was fast running out of money. So in 1965, for the first time since it was set up, the family firm went public. It was careful to keep control (various family members made up seven of the nine-strong Board) and the influx of cash was spent on new equipment and a big advertising campaign. But it was only putting off the inevitable. Within a couple of years, the company was back to square one and ready to accept that a takeover was the only way to survive.

There were two suitors. Bangor Punta was keen, but had a reputation as an asset-stripper. The American Machine and Foundry Company, on the other hand, had a strong engineering background; it promised that Harley's existing management would keep control, and chairman Rodney C. Gott was a Harley owner! After some courtroom wrangling, Harley-Davidson shareholders accepted the deal, and Harley-Davidson became a subsidiary of AMF in January 1969. It was a family firm no more.

TEXACO

OPPOSITE: All the bikes pictured here are Shovelhead Electra Glides, which sold steadily right through to the late 1960s and into the 1970s. Although by now they are in some ways outdated, for many American riders they remain the definitive touring bike. On the far left is an early 1966 version, the first year for the updated Shovelhead V-twin, while on the right are two examples showing restrained and fully loaded approaches to the 'dresser'.

RIGHT: A Shovelhead Electra Glide.

1970s: Have We Got Trouble!

When Willie G. Davidson styled the Super Glide in 1969, he may not have known that he was opening up a whole new era of Harley history. The Super Glide was a milestone bike for the company, even though it was no bigger or faster or more advanced than any previous Harley-Davidson (let alone any other motorcycle). What it did do was to mix and match existing components to produce something new. Or at least, something that *looked* new. This worked again with the Low Rider in 1977, and enabled Harley-Davidson to get through some very difficult times when it couldn't afford, or didn't have time, to introduce genuinely new bikes.

ABOVE: The 1971 FX Super Glide, with original 'boat tail' rear end. Most of these were junked and conventional rear ends substituted.

OPPOSITE LEFT: Engine detail of the FXE Super Glide. The 'E' denoted electric start, otherwise it was identical to the FX.

OPPOSITE RIGHT: A Super Glide with its later rear mudguard which allowed them to sell better.

Grandson of the original William Davidson, Willie G. had joined the family firm back in 1963 after working as a designer for Ford and Brooks Stevens. He was a professionally trained designer who happened to have grown up with motorcycles, so he was ideally placed to interpret what Harley riders wanted, and put it into practice. And the Super Glide was a perfect example of what he could do. Harley-Davidson had long tried to distance itself from the custom movement because of its associations with the outlaw element. But the reality was that thousands of Harley riders were modifying their bikes in this way, and it was a golden opportunity for Milwaukee to produce a ready-made custom bike. The Super Glide was therefore the first factory custom. The idea was simple: Harley married the frame, engine and gearbox of the touring FL with the lighter-weight forks, 19-inch (48-cm) front wheel and small headlamp of the XLCH Sportster. Despite the incongruous 'boat tail' glass fibre rear end that most owners junked, it looked long, low and handsome, and opened up a new market for the customizers who wanted an off-the-peg custom bike.

Not that the rest of the range was neglected. The XLCH Sportster was opened out to 997cc (the return of the classic 61ci engine size) though there was no attempt to supertune it. Milwaukee had accepted that it could never keep up in the horsepower race, and it was pointless to try. The buyers evidently weren't worried, and bought 20,000 Sportsters in 1973.

The small imports from Aermacchi were more of a problem. The Sprint, which had sold well for a while, had long since been overtaken by more sophisticated Japanese lightweights, and Harley-Davidson's answer was to meet these head-on with a range of two-stroke trail bikes.

Their origins lay in the Rapido, a 7-ci (125-cc) commuter bike that was unveiled in 1968. But as the TX125, it wore trendy off-road clothes, knobbly tyres and oil injection. An 11-ci (175-cc) version soon followed, and a 15-ci (250-cc) in 1975. On paper, they looked the business: oil injection two-strokes with CDI ignition, five-speed gearbox and authentic trail bike looks. And for a while they sold well with over 11,000 250s alone in 1975; but it was short-lived success. It was a measure of how fast the Japanese were progressing that the Harley-Davidson-badged trail bikes were looking a little crude and unfinished by 1976. Question marks remained over Aermacchi quality, and in 1978 Harley-Davidson pulled out of Italy altogether, selling the Varese factory to Cagiva. It has never sold a small motorcycle to the public since.

So what was AMF, the new owner of Harley-Davidson, up to all this time? It had a clear goal: the motorcycle market was booming, but Harley-Davidson wasn't making many bikes. The AMF plan was to boost production dramatically, thus generating the money needed to fund much-needed new models. They certainly succeeded in boosting production, moving bike assembly to York, Pennsylvania, and more than doubling the output of engines and gearboxes at Juneau Avenue in three years. But, as Rodny Gott later admitted, it was all pushed through too quickly with little regard to quality. As a result, AMF-era Harleys soon gained a reputation for poor quality, while a surfeit of 'top-down' management meant that the long-term Harley-Davidson employees were being ignored.

Things weren't going as planned, so AMF man Ray Tritten made a thorough study of the whole business, finding a lack of professionalism (in both engineering and marketing), complacency with regard

to the Japanese, and inefficient production.

With characteristic energy, Tritten set about cutting costs and introducing proper forecasting. He also brought in two men who were to play key roles at Harley-Davidson over the next 20 years. Jeff Bleustein was a fomer associate professor of engineering at Yale University and managed to transform the engineering department by bringing in more professionals. Vaughn Beals (already an AMF man) was put directly in charge of Milwaukee, and one of the first things he did was to sort out a long-term new model programme.

In the short term, the V-twin would be updated to see Harley-Davidson through the next few years, which appeared in the form of the Evolution motor. But for the future (and this was the radical part) there would be an all-new family of engines, 30–80ci (500–1300cc), two cylinders to

THIS PAGE: A rare line-up of original 1971/72 Super Glides. Front drum brakes were still relied upon to stop these heavy motorcycles, but disc brakes were on the way.

OPPOSITE: A variety of Shovelhead Electra Glides, all showing the single front disc brake that Harley adopted in the early 1970s. The bike at top left has a sidecar attached which could still be ordered from the factory, though not many took advantage of the option.

six (the NOVA programme). The plan was clear, a product-led recovery, funded by AMF.

But even with NOVA and Evolution going full steam ahead, Harley-Davidson still needed to survive in the meantime. Once again, Willie G. came to the rescue with two new bikes, both of them based on the Super Glide principle of mix 'n' match existing parts. One was a flop, the other a huge success. The XLCR Café Racer, with its vaguely European looks and all-black colour scheme, was just too far out from the traditional Harley look to be acceptable. It was dropped after a couple of years. The Low Rider, on the other hand, took the concept of a factory custom even further than the Super Glide. This was much closer to a real chopper, with lengthened and raked forks, low-riding 27-inch (69-cm) seat height with a king and queen seat. It was a canny blend of old and new, with tank-mounted instruments and 1903-style tank script, but with a twin disc front end and alloy wheels as well. The buying public loved it, and the Low Rider became the best- selling Harley (basic Sportster apart) over the next few years.

This was all very well, but AMF was growing impatient. For over ten years, it had been putting a great deal of money into Harley-Davidson, for very little return. Although there has been a tendency to castigate AMF as faceless big business, it really did want to make a success of Harley-Davidson. But after ten years of effort, the company was still only making small profits, and certainly not enough to fund the NOVA programme, which according to a team of consultants would need $70 million to reach production. There was a time when AMF would have signed the cheque, but not any more. Vaughn Beals made the recommendation; it was time to sell.

1980s: Like a Japanese Miracle

It was the best of times, it was the worst of times. For Harley-Davidson, the 1980s brought near-bankruptcy, a dramatic turnaround and a new-found prosperity that it still enjoys today. Not since the 1930s had it come so close to a complete close, and never had it experienced such a dramatic reversal in its fortunes. The decade had an inauspicious start: Harley's share of the US big bike market had slumped to one-third by 1980, and just 23.3 per cent by 1983. In that year, for the first time ever, it sold fewer big machines than Honda. Yet in 2000, a revitalized Harley-Davidson overtook Honda again, not only in the big bike market, but overall. For a company that until recently sold nothing smaller than the 54ci (883cc) Sportster (Honda of course, offers a complete range of bikes) it was an astonishing achievement.

But the seeds of this remarkable revival were being sown even in the darkest days of the early 1980s. Vaughn Beals' vision of a product-led recovery was showing signs of life, thanks to the revitalized engineering department led by Jeff Bleustein. First off, the FLT, launched in 1980, was almost an entirely new motorcycle, and was the first sign that Harley-Davidson was once again taking R&D (Research and Development) seriously.

The new FLT Tour Glide was Harley's response to Honda's GoldWing, a smooth and sophisticated tourer that had been steadily eating into the company's traditional market. To meet this formidable competition, the FLT had a new steel backbone frame, which rubber-mounted the venerable Shovelhead V-twin (just about the only part of the bike that *wasn't* new).

Rubber block mounts at the front and the swinging-arm pivots, plus adjustable locating links at the front of the engine and the top, were master strokes. They allowed vibration through when idling, but virtually eliminated it when the bike was on the move. One magazine road test likened it to the gentle thrumming one might feel through the deck of a steamship. It was a huge step forward, which has since been applied to nearly every Harley model and played a big part in increasing their acceptability to non-Harley riders.

The bike's geometry saw innovative thinking too: the forks were mounted off-set to the steering head, which allowed a steepish steering-head angle (for quicker steering) with long trail (for stability). There were a few changes to the Shovelhead in the form of electronic ignition and a quieter exhaust system, while the gearbox acquired a fifth speed. The twin headlamp fairing was new, as was the twin disc front end, the enclosed

All the bikes on this page and opposite demonstrate how the Evolution V-twin spread right through the range in the 1980s, and made possible Harley's product-led recovery. OPPOSITE is a Softail Custom with rear suspension hidden for that hardtail look. LEFT are two Dyna Super Glides, the lower one showing various accessories. ABOVE is a fuel-injected FLHTCUI Electra Glide.

chain and 5-gallon (22-litre) fuel tank. The FLT was really another of Harley's milestone bikes, and is still part of the range, 20 years on.

Later that same year came the Sturgis, which introduced another innovation – belt drive. It was the first production motorcycle to offer both primary and final drive by toothed belts, and the advantages were many. Belts need no lubeing and have a 20,000-mile (3220-km) life; they are quieter and smooth out the drive; they allow a dry (and thus leak-free) chain case, which gives the clutch an easier time. So successful was the system that, like the rubber-mounted motor, it has since been applied to every Harley-Davidson you can buy.

It is odd, however, that the new belt-equipped FXB-80 Sturgis was not designed as a tourer (surely the first market for a clean, long-lasting final drive) but was the

Low Rider. And in every other respect, this was a traditional Harley, with solid-mounted motor and a four-speed gearbox: sales though, convinced Harley-Davidson that belts were the way to go.

This was all very well, but the company itself was in deep trouble. AMF wanted out, and Vaughn Beals actually encouraged the move by recommending that his employer sell its troublesome subsidiary outright. And when AMF management agreed that this was a good idea, it just so happened that Beals had a management buyout proposal ready and waiting. You had to admire the man's subtlety.

The negotiations dragged on into 1981, but eventually, with the help of some massive bank loans, Harley-Davidson's management bought the company back for $80 million. To celebrate, some of them (including Willie G.) rode from York to Milwaukee, calling at jubilant dealers on the way. Once more, the advertisements ran, 'The Eagle Soars Alone'.

But they weren't out of the woods yet: market share was still slipping and quality was not what it should be. New products were on the way, but the company itself had to become more efficient. This happened by adopting Japanese business methods of which there were three in particular: just-in-time inventory cut-down on money tied up in stocks of parts; employee involvement using the skills of people on the shop floor to affect decision making; and every worker was given the means to monitor his or her own quality and output. All three, known as the 'Productivity Triad', had a tremendous effect on efficiency, productivity and quality, the problems which had bugged Harley for decades. More good news came in 1983, when Harley-Davidson's third application for tariff protection against large imported bikes was finally granted; it was for five years only, but gave the company a crucial breathing space.

But for most Harley riders, this all happened behind the scenes; what really marked the 1980s was the Evolution twin. As the name implied, it wasn't all-new (a new top-end on the old Knucklehead-based crankcase) but was as much of a leap forward as the Knucklehead had been nearly 50 years before. Aluminium-alloy cylinders were lighter and ran cooler, a narrower valve angle allowed shorter, straighter ports, the valve train was lighter, the con-rods stronger and tolerances all-round were tightened up dramatically. The result was more power than the old twin, and it was quieter, more reliable and didn't leak oil. Best of all, Harley-Davidson had finally got a new engine right first time.

It cannot be stressed enough that the Evolution V-twin saved Harley-Davidson, and made its 1980s turnaround possible.The other new model in 1983 was less noticeable, but just as significant. For once Bill Davis had built a Harley that cleverly hid its rear suspension units under the gearbox resulting in hardtail looks, but with at least a little rear suspension movement. Vaughn Beals (so the story goes) saw it, liked it and bought the patent. Since then, the Softail, as it was called, has been a hugely successful part of Harley-Davidson's range, combining the looks of a 1940s or '50s Hydra-Glide with a modicum of ride comfort. The Springer Softail of 1989 took that to its logical conclusion, combining the Softail rear end with a redesigned version of the Springer forks that Bill Harley had initiated in 1907. This was the ultimate retro machine.

But even as all these good things were happening, the company was facing collapse and rebellion. In 1985 Citicorp, one of the main financers of the buyout, announced that it wanted its money back, which would have meant bankruptcy. Only days before the plug was pulled, Richard Teerlink, Vaughn Beals and Tom Gelb managed to get new funds from Heller Financial. Meanwhile, long-suffering Harley dealers had become so disgruntled with the new regime (which insisted on lower margins, and new investment in training and marketing) that they formed their own independent association. Gradually, compromises were made and the rifts were healed.

Just as Evolution transformed the big twins, so it did the same for the Sportster. A 54-ci (883-cc) Evo Sportster replaced the old one-litre Shovelhead version in 1984, with a 67-ci (1100-cc XLH following on soon after (soon boosted to the XLH1200 we know now). Still with a four-speed gearbox and chain drive, Harley emphasized the little 883 as a loss-leader entry model. Even now, it undercuts some Japanese cruisers, though even in Evo form, it's slower and less sophisticated than any of them. If the 883 was a sensible cut-price way of boosting sales, the XR1000 did the opposite. Its engine, using modified XR750 parts, was expensively hand-assembled, but the handsome result was still a long way behind Japanese sports bikes. On the other hand, it did form the basis fo the first Buell, which led to greater things.

Only a year after its eleventh-hour flirtation with financial collapse, Harley-Davidson was visibly on the road to recovery. The Evolution had been well received, and sales were booming. It was an ideal time to go public and to raise money by selling shares on the general market. The issue was such a succes that $90 million was raised, enabling Harley to pay off many of its debts with enough left over to buy the motorhome maker Holiday Rambler. In

In the late 1980s/early '90s, Harley-Davidson nostalgia reached new heights. OPPOSITE: The Springer Softail marked the return of the springer fork. BELOW: The Heritage Softail Nostalgia, in all its black, white and chrome glory.

theory, it could have used the money to resurrect the NOVA engine programme, but by then it was clear that the Evolution (originally intended as a stop-gap) would see the company through many years to come: in any case, the buying public expected a Harley-Davidson to have an air-cooled V-twin engine, not a water-cooled four.

By the end of the 1980s, the recovery was complete. The company made a $17.7 million profit in 1987 and $27 million the year after. In fact, it is instructive to compare the beginning of the 1980s to the end: productivity up by half, US market share doubled; inventory cut by 75 per cent, scrap/rework down by two-thirds and an annual profit of $59 million. Police business (once highly prized, and lost in the bad old days of poor quality) began to return and exports increased dramatically; not only that, but it must have caused some satisfaction at Milwaukee when Japan became its foremost overseas market. When Ronald Reagan visited the York plant, it was the ultimate seal of official approval; Harley-Davidson had become the American success story of the 1980s.

1990s and into the Next Century: Ever Upward

If the 1980s was Harley-Davidson's turnaround decade, then the 1990s saw its consolidation. The Harley bandwagon, which set off in the mid-1980s and gathered pace in the early '90s, showed no sign of stopping in 2000, or even slowing down a little. It was a measure of how far the company had come that in 1990 it could call a motorcycle 'Fat Boy' and be taken seriously. The market had changed, and there was a new breed of affluent bikers for whom motorcycling was a weekend hobby rather than a means of transport. For them, it hardly mattered that the Harley was slower, cruder and more expensive than the Japanese opposition. What did matter was that it was way ahead of all of them on high-profile imaging – a solid no-nonsense all-American motorcycle.

ABOVE: The Evo Fat Boy: who else but Harley-Davidson could get away with a name like that?

OPPOSITE: The Road King was a kind of stripped-down Electra Glide. This is a 2000 model with Twin Cam 88 engine.

The way that Harleys developed through the 1990s reflected this, with more emphasis on styling than engineering. Of course, there have been engineering advances like fuel injection, balance shafts and sealed wheel bearings, but the company has chosen not to invest in anything radically new, even though, perhaps for the first time, it could easily afford to. There were rumours of a 95ci (1560cc) version of the Evo in 1989/90, an attempt to overtake the 85ci (1400cc) Japanese lookalikes. It came to nothing, as the bigger engine reportedly had unacceptable bore and piston wear. That was something else that was different about Harley-Davidson – it now knew better than to launch anything substandard.

The Fat Boy summed all this up. It was the nearest thing to a new Harley model in 1990, though really no more than a Heritage Softail with a few styling tweaks and a new paint job. Frame, tank, mudguards – almost everything was finished in silver, with the 'fat' look emphasized by solid wheels, wide and heavy valanced mudguards and wide tyres. All with a solidly-mounted engine, vibration and all, because that was what some riders wanted. An interesting contrast was the FXRT, also current in 1990. This had a rubber-mounted motor, proper modern (not stylized) fairing, air-adjustable forks and twin front disc brakes. The FXRP police version went down well, but Harley-Davidson sold a mere 304 civilian FXRTs in 1990, and over 4,000 Fat Boys. You couldn't have had a clearer indication of the reasons why people were buying Harleys. Then there was the 'Harleywood' phenomenon in which a new Harley-Davidson became the accessory of choice

LEFT: Top of the range in the early 1990s was the Ultra Classic Electra Glide.

ABOVE: The Sportster was the quiet success story of the 1980s and '90s, bringing many new riders into the fold, especially in its updated five-speed Evo 883 form.

for Hollywood names. Sylvestor Stallone, Cher, Bruce Willis and countless others were happy to be seen in the saddle of a Harley and millionaire Malcolm Forbes even formed his own club for similarly wealthy Harley riders. One less affluent owner wryly observed that in Hollywood, Harley-Davidsons had become like a certain part of the human anatomy – everyone had one!

Meanwhile, certain advances did trickle down through the range. In 1993, the 883 Sportster finally acquired belt-drive and a five-speed gearbox, both long overdue. It was still relatively cheap to buy, undercutting even some of the equivalent Japanese cruisers and played a crucial role in attracting new riders into the Harley fold. It was still doing that in 2000, seven years on from belt-drive, and 16 from the 883 Evolution motor introduction. It also retained a charming simplicity that was

lacking in some of the bigger, more expensive bikes. The Dyna Glide of 1993, on the other hand, had a new frame, but paradoxically the sole reason was to make it look like an old one! The bike was an update of the mid-range rubber-mounted FX series, but the idea here was to emulate the 1970s Low Rider, the engine of which kept its rubber mounts, but were cunningly hidden by the new frame. The names also evoked nostalgia, and the first Dyna Glide was a limited-edition Sturgis (remember 1980), followed by the raked-fork Wide Glide. Mechanically, the Dyna stuck with the 82ci (1340cc) Evolution engine, now in its tenth year and still in a very mild state of tune.

Harley-Davidson's foray into the law courts in the mid-1990s underlined how important its image had become. With the Japanese making big, reliable, good-looking V-twins, Harley became increasingly

ABOVE LEFT and RIGHT: Screamin' Eagle tuning parts could be applied to any Harley. This is a limited-edition Twin Cam Tour Glide so equipped.

LEFT: A 2000 Twin Cam Dyna Glide but, at a quick glance, it could be taken for a machine built 30 years ago.

concerned to protect those things that made it unique. The names 'Glide', 'Sportster' and 'Hog' were all registered as trademarks, and even the distinctive 'potato-potato' exhaust note (a product of the Harleys' 45-degree cylinder and a common crankpin for both con-rods) was subject to a patent application. 'The Harley-Davidson sound functions as a (trade) market and identifies Harley-Davidson alone as the source of the goods emitting that sound.' The company actually withdrew that application after a few years, but the fact that it attempted at all underlines the fact that these bikes weren't being bought for their A to B transport capability.

So strong was the nostalgia element that the touring FLs had been losing sales to the more obviously retro Heritage Softails. So Harley unveiled the FL Road King, a stripped-down tourer that harked back to the original Electra Glide, with no fairing and a detachable pillion seat. There were signs, though, that Harley-Davidson realized that all this wasn't sufficient on its own.

In 1995, fuel injection appeared on the US/Canadian-market Electra Glide Ultra, with electronic engine control according to temperature, load and so on. It didn't make the bike go any faster, but it did improve economy and cold starting as well as emissions. Even sporting pretensions were creeping back into the range: in 1998, the Sportster 1200 Sport gained more power from twin-plug heads, a higher compression and new cams to add to the twin front discs and adjustable suspension it already had.

But it was the final year of the century that saw Harley's biggest news for a long time. The new Twin Cam 88 motor was unveiled, at first for certain models only, but clearly intended to replace the

Evolution in time. At 88 cubic inches (1450cc), hence the name) it was the largest production Harley engine yet, though the company had evidently decided it wasn't going to get into a cubic-inch war. Victory (92ci/1507cc), Yamaha (98ci/1602cc) and Honda (110ci/1800cc) all announced even bigger V-twins soon afterwards.

Despite its name, the Twin Cam didn't have twin overhead cams, but two cams on the shaft; in fact, it was still air-cooled and certainly contained no radical changes. But although it looked similar to the Evo, the Twin Cam was very different inside, with a bigger bore/shorter stroke, reshaped combustion chamber, new ignition and lubrication systems, oil-cooled pistons, redesigned breathing and die-cast crankcases. It was claimed to be stiffer, stronger, more powerful and more oil-tight than the old engine. Harley-Davidson also claimed that only 21 parts had been carried over from the Evolution, 'most of them screws', according to the brochure. The Twin Cam was still rubber-mounted, but for 2000 the 88B was launched, with twin contra-rotating balance shafts for what must be the smoothest Harleys yet.

So as the 21st century got underway, Harley-Davidson was facing it with a new confidence after 15 years of almost uninterrupted growth. There was record production, record sales and record profits and it seems a fair bet that it will reach the 100th anniversary of Bill Harley and Arthur Davidson's first prototype in good shape.

ABOVE: The latest incarnation of the Electra Glide, with the 88-ci (1450-cc) Twin Cam motor.

RIGHT: For those who wanted a slightly more contemporary tourer, the Tour Glide was still on offer in 2000.

2008/09 Night Rod.

Klaus Naur/Flickr

2008 Road King

Johnny Jupiter Photo's/Flickr

A–Z OF HARLEY-DAVIDSON

Abresh Body Shop
Abresh, based in Milwaukee, built sidecars for Harley-Davidson for many years, starting with sidecar and van bodies made to special order. In 1936 it introduced an all-new steel-bodied sidecar, and carried on producing steel sidecars for the factory right up to 1966. The following year, Harley moved over to fibreglass bodies, laid up in the former boat-building facility at its Tomahawk plant.

Accessory Groups
To reduce costs and rationalize the optional extras business in the 1930s, Harley-Davidson reorganized its accessories into Standard or Deluxe groups, which were bought as a pack. For 1936, for example, the Standard group for the side-valve twins was a Safety Guard (engine crash bars), a side stand, a lighted speedometer with maximum-speed-attained hand, Ride Control and a steering damper. If you could afford the Deluxe group, it included dice heads on the switch panel, saddlebags and chrome plating for the fender lamp, stoplight and dice shift knob. In 1937, 90 per cent of new machines were ordered with one or the other of these groups.

Prest-o-Lite (press-to-light) reservoir for acetylene lighting. Like other manufacturers, Harley-Davidson moved to electric lighting as soon as possible.

Acetylene Lighting
Before charging systems and batteries became reliable enough to make electric lighting universal, motorcycles (Harleys included) used acetylene lights. To switch them on (though 'switch' probably isn't the right word), you turned on a tap, which dripped water onto carbide pellets which gave off carbide gas, which could be lit. The result wasn't searchlight power, but was better than nothing. In fact, a book entitled the *ABC of the Motorcycle* (published in 1910) states: 'There is only one lamp to use on a motorcycle and that is the acetylene ... Electric lights such as are used on automobiles are out of the question as they require too much extra power over that generated by the magneto.' The author was wrong: Harley-Davidson was offering electric lights on the J-model in 1915, and by 1926 all the road-going twins were so equipped.

Advertising
Harley-Davidson's advertising has become something of an art form in itself. The Milwaukee advertising firm of Klaus, Van Peterson and Dunlap obtained the Harley-Davidson contract in 1907, and incredibly, kept it for the next 50 years. Not until well after the AMF takeover was there a serious change in advertising policy, but when there was, it caused a serious rift with AMF. In 1976, AMF hired the prestigious Chicago advertising agency, Benton & Bowles, for all its recreational subsidiaries, of which Harley-Davidson was one. All

The Aermacchi era: Harley-Davidson's attempt to compete with the Japanese through the 1960s and '70s was via the Italian-made Aermacchis, which were fitted with H-D badges and sold in North America through H-D dealers. Not all dealers were pleased with this development, deriding the bikes as 'not real Harleys', which was true enough. However, they did give Harley a toehold in the small bike market, and thousands were sold each year. Whether this was worth the company's while in the long term is less clear: there were serious problems of communication with the Italian factory and quality control was a permanent headache. LEFT is a period 1970s advertisement for the 11 and 15-ci (175 and 250-cc) two-strokes, the last Aermacchis to be sold as Harley-Davidsons. Below is the neat four-stroke engine unit of the 250 Sprint, the first Aermacchi to wear an H-D badge.

Harley-Davidson national advertisements now had to adhere to strict guidelines. The trouble was that B & B had no experience of the motorcycle market, and the advertisements it produced were unsuitable. Harley-Davidson got around this by expanding other forms of publicity, such as dealer co-operative schemes, and direct mail. This was not popular with AMF, particularly when Harley-Davidson placed a full-page ad in *Easyriders* magazine for, at the time, AMF wanted to distance itself from the 'outlaw' end of the market. In the end, the parent company relented, and put Harley-Davidson back in charge of its own advertising budget.

Aermacchi

The Italian motorcycle maker, bought by Harley in 1960, and sold in 1978. Harley's reason for buying into Aermacchi was simple: by the late 1950s, it was clear the company needed to get into small motorcycles as British imports, the Triumphs, Nortons and BSAs, and to a lesser extent the Italians, were dominating the market. Although Honda had yet to secure its 'Nicest People' revolution, small motorcycles were increasingly popular, and Harley was losing market share as a result. All it could offer was the outdated 7ci (125cc) Hummer, and the quickest solution was to buy someone else's small bikes and stick Harley-Davidson badges onto them.

Aermacchi seemed like the obvious partner: it had been an aircraft manufacturer before the Second World War, but turned its attention to motorcycles in 1954. The first prototype was actually electrically-powered, but a 125cc machine, the 125N, soon followed. It was really a cross between a bike and a scooter, with a step-through fame (scooter-style) and motorcycle-type dummy fuel tank. Others followed, notably the futuristic looking 10.5-ci (172-cc) Chimera in the mid-50s, while in competition Aermacchi secured new 50cc and 75cc speed records.

But the company was small and in a weak position, despite gradually improving sales of its 11 and 15ci (175 and 250cc) four-stroke singles, with their characteristic horizontal cylinders. This was what attracted Harley: the Aermacchi singles were modern and lightweight for their time, but the Italians needed investment to carry on making them. A deal was struck: Harley-Davidson bought the whole caboodle, and Aermacchi began exporting the singles to America, complete with Harley-Davidson badges. They were successful, and despite fierce competition from the Japanese continued to sell right into the 1970s, though were looking increasingly outdated by then, despite the later addition of electric start.

But Aermacchi was not standing still, and 1967 saw the introduction of the M125

Rapido, which formed the basis of a whole new generation of two-strokes. Off-road bikes of 5 and 6ci (90 and 100cc) followed, and the much-updated TX125 trail bike in 1973. A 175 version followed, then a 250, which was fast but more expensive than the opposition. In fact, the pace of development was so fast that Aermacchi just could not keep up with the Japanese, and sales began to drop off. In 1978, faced with its own problems at home, Harley-Davidson sold the Aermacchi concern to Cagiva, a fledgling Italian bike maker which utilized the factory to good effect.

AFL-CIO
Awarded the Union Label to Harley-Davidson and four trade unions in 1988, in recognition of 'the foresight and co-operation shown by unions and management in helping to revitalize the company'. The Allied Industrial Workers Union and Machinists and Aerospace Workers shared the award.

Aircraft
There have been a couple of attempts to produce Harley-Davidson-powered light aircraft. Harvey Mummert worked for the Curtiss-Wright Aviation Company in the early 1920s, and happened to be a Harley-Davidson enthusiast. He built a wooden-framed monoplane, powered by a J-model 61ci (1000cc) V-twin. Lightly modified V-twins usually shook light aircraft to pieces due to the vibration, but Mummert had balanced his for its 1,600–1,800rpm running speed and the plane worked well, taking second place in the National Air Race. Harvey Mummert later built a small racing biplane for Sarter Tiffany, this time with the J-twin inverted and which also competed successfully.

Thereafter, most light aircraft with motorcycle engines opted for the Henderson or Indian fours, with their inherently lower vibration. An exception was the one built in the early 1930s by plane enthusiast Lester Long, who built up a flat-twin motor for aircraft from J-model cylinder assemblies and a new crankcase. The resulting 80ci (1310cc) unit was well balanced and produced 30hp, able to spin a 5ft 2-in (1.575-m) propellor at 2,750rpm. Long produced 50 of the engines himself, after Walter Davidson had shown no interest in producing it in-house.

Albrecht, Paul
A Harley-Davidson competition rider in the 1950s.

Alexander, Ronnie
A racing mechanic who worked on Ray Calborn's XR750s in the early 1970s. He was later recruited to the company's quality audit department, which had its work cut out from the late 1970s after years of quality being a poor relation to production.

Allchin, T.R.S.
A prominent English Harley-Davidson rider in the 1920s.

Allen, Irwin D.
President of the Federation of American Motorcyclists, which was voluntarily dissolved in 1919 because of declining membership and an inability to control motorcycle sport. It was replaced by the American Motorcycle Association (AMA).

Allied Industrial Workers Union
Major trade union at Milwaukee, representing production and maintenance workers.

Allis-Chalmers
Harley did much subcontracting work for this well-known maker of tractors and other agricultural machinery during the Second World War. Among the more interesting projects was a powered snow-sledge, a suitable engine being the side-valve flat-twin XA motor that Harley-Davidson was developing during the war. This itself was heavily based on the German BMW and the XA motorcycle (of which a few were built for the military) and came complete with shaft drive. Two XA power units, with three-speed and reverse gearbox, were supplied to Allis-Chalmers, but no more was heard of the project.

Alzina, Hap
An Indian dealer in Oakland, California during the 1920s. He later became Indian's west-coast distributor and was so successful that, according to author Harry Sucher, it was Alzina's sales and profits that kept Indian afloat during the difficult years from the late 1920s to the early '30s.

AMA (American Motorcycle Association)
America's governing body for motorcycle sport and manufacturing. It was founded in 1920 after the pioneering Federation of American Motorcyclists (FAM) was voluntarily wound up. Although originally intended as an independent democratic organization, the AMA later became dominated by trade interests. An example was the institution of the 21 cubic-inch racing class in the 1920s, which just happened to favour Harley-Davidson's new 21ci (350cc) single. The same thing happened in the early 1970s, when the AMA changed flat-track rules to allow overhead-valve 750s to compete and Harley's XR750 fitted the bill. What the AMA did achieve, however, was proper organization of motorcycle sport in America, which had previously been lacking.

Americade
Previously the Aspencade Motorcyclists Convention, now renamed and moved to the town of Lake George, New York. Unlike the custom-orientated Sturgis and Daytona, this one is for touring bikes, though like them, it is Harley-Davidson-dominated.

American Association of Motorcycle Road Racers
A group of racers in the 1960s which, increasingly dissatisfied with the AMA's restriction of non-approved events, formed its own organization. When big names, like Dick Mann, began to compete in these 'unofficial' events, it soon became clear that the AMA would have to move with the times and accept different forms of competition that were becoming increasingly popular with the public. It was in these exchanges, often through the pages of *Cycle World* magazine, that the AMA was increasingly accused of being dominated by Harley-Davidson and not representative of American motorcycling in general.

American Car & Foundry
The original name of the American Machine and Foundry Company (AMF), which took over Harley-Davidson in 1969. It started off in the railway rolling stock

business, and diversified into yachts, bakery and tobacco, among other things.

American Machine & Foundry Company (AMF)
AMF was the huge engineering concern that took over Harley-Davidson in 1969, selling it back to the management 12 years later. It was a big firm, based in engineering but with a desire to expand the leisure side of its business: moreover, chairman Rodney Gott was a Harley rider! In the event, AMF forestalled a hostile bid for Harley-Davidson by Bangor Punta. Since then, AMF has not had a good press, being blamed for everything that went wrong at Milwaukee in the 1970s. AMF-era Harleys were for years seen as a real low point, regarded as bad bikes produced by a profit-motivated company that had no knowledge of or interest in motorcycles. It's an unfair assessment. While it is true that AMF was motivated by the growing motorcycle market, the Harley-Davidson it acquired was under-capitalized, with outdated plant producing too few bikes. So AMF spent a lot of money re-equipping Harley's Juneau Avenue factory (built in the 1920s) and moved motorcycle assembly to its own plant at York, Pennsylvania. It also began R&D work on a radically updated V-twin and an all-new range of water-cooled machines. None of this was the work of a heartless asset-stripper.

The problem was, in its anxiety to make the best of a booming bike market and see a return on its investment, the company forced up production rates much too fast: Harley-Davidson built 27,000 bikes in 1969, and 60,000 three years later, and this was dictated from the top, with little regard to quality. Although there were assurances that moving production away from Milwaukee would not mean redundancies, some people did, however, lose their jobs. The end result was a disgruntled workforce, frustrated management and sceptical dealers, left to cope with the poor quality machines that AMF was expecting to sell. But in the final analysis, despite the mistakes AMF made and its heavy-handed approach to reorganization, there is a good case for saying that it actually saved Harley-Davidson. The company stood little chance of surviving on its own into the 1970s, having run out of money: it had nowhere else to go.

American Motorcycle Manufacturers Association
The first organization for US motorcycle makers, founded in 1914, with Arthur Davidson on the board of directors. It was instrumental in the setting up of the AMA when, at the 1919 annual meeting, it was moved that the manufacturers take over the responsibilities of FAM, which they did. This formed the nucleus of the AMA. It was later renamed the American Motorcycle and Allied Trades Association.

American Stock Exchange
For a company approaching its 100th birthday, Harley-Davidson has had very little to do with stocks and shares. The reason is simple. For the first 60 years, it was owned by the Harley and Davidson families and only went public in 1965, when in dire need of fresh capital. AMF took over the whole thing in 1969, selling the company back to the management team in 1981. Five years later, to raise more capital, Harley-Davidson finally went public again, and was floated on the American Stock Exchange; in other words, anyone with enough funds could buy a few Harley shares, and lots of people did – not

When rebadging Aermacchis as Harleys and floating itself on the Stock Market wasn't enough to keep the company going, Harley-Davidson bowed to the inevitable and sold out to the American Machine & Foundry Company. The years that followed would be seen by many as a disaster, with over-production of poor quality machines by an ownership which didn't understand the motorcycle. The reality wasn't quite as simple, and there's a case for saying that Harley-Davidson would not have survived the 1970s without the AMF takeover. For better or worse, AMF logos began to appear on Harley-Davidson fuel tanks (LEFT) and in advertisements (BELOW). However, when the company was eventually bought out by its own management in 1981, these logos were the first to go.

only traditional investors, but Harley-Davidson dealers, employees and riders. Some later admitted that they intended to sell the shares at a profit (they nearly tripled in value in four years) and buy a bike with the proceeds!

Amtorg Company
Amtorg was established by the Soviet Union and based in Washington during the Second World War, and one of its tasks was to source motorcycles and send them back to Russia. A Mr Karsov was in charge of motorcycle buying, and had bought some Indians in early 1940. He'd been doing the same job back in 1924, when Charles Cartwright, Harley-Davidson's salesman in New England, had helped him buy some Harleys on credit. It was no surprise then, when the two men met again, that the Soviet Union switched back to Harleys.

Andres, Brad
Brad Andres wasn't a factory-backed racer, but achieved great things with the side-valve KR in the 1950s. He won the National Championship in 1955 and was second to Joe Leonard the following year. The KR itself was something of a minor miracle. Despite an apparently antiquated side-valve motor (with roots going back before the Second World War), it won the American flat-track National Championship 13 times.

Andres, Leonard
Father of Brad Andres, Leonard was a Harley-Davidson dealer and tuner who took the young Ray Calborn under his wing and encouraged his early career.

Antique Motorcycle Club of America
The club caters for machines over 30 years old, and anyone interested in old bikes can join. Contact Dick Winger, Antique Motorcycle Club of America, 2008 Fitchburg Rd, Stockbridge, Massachusetts 49285.

In the context of American racing, and like its predecessor, the WR, the KR racer was developed into a very competitive bike, drawing on a couple of generations of side-valve tuning know-how. ABOVE: Notice how the bike has been made radically lighter.

Appeal Manufacturing & Jobbing Company
Held the Harley-Davidson dealer franchise in Los Angeles from 1918, but was soon replaced. The problem was that the south-west and far-west of America were developing into very significant markets, making up half of all domestic

The Antique Motorcycle Club of America's national show at Davenport. Harleys, like this Duo Glide, naturally feature prominently among club members.

sales. Distances were great, and there were many enthusiasts in the area. Unfortunately, Appeal Manufacturing lacked ambition: it refused to advertise, there was no planned sales campaign and certainly no hard, aggressive sales techniques. The result was that Indian, Excelsior and other makes sold far more bikes than Harley-Davidson.

Arthur Davidson, then sales manager, acted quickly, persuading Rich Budelier, who had made a great success of selling Harleys in San Diego, to take on the Los Angeles franchise. Another factor in boosting Harley-Davidson sales in this important market was the decision to hold stocks of machines in warehouses in Los Angeles and San Diego. Previously, Western dealers had to order specific bikes from Milwaukee, and wait.

Appel, George
Appel worked for electrical components maker Remy in 1919, and it is said that he was hired by Harley-Davidson to sort out the problems with its own electrical components. At first, the electrically equipped J-model from 1915 used the Remy generator, lighting and horn, but Milwaukee introduced its own electrical equipment on the Model 20 J in the summer of 1919. The official reason was that Remy components were no longer up to Harley standards: in reality, the in-house equipment had plentiful problems of its own, and some dealers ended up replacing it with Remy parts.

Appelle, Harvey
A senior loan officer with Citicorp Industrial Credit, the company which largely financed the management buyout of Harley-Davidson from AMF in 1981. An early meeting between Appelle, Harley-Davidson chairman Vaughn Beals and Gary Ward (the AMF man in charge of Harley-Davidson's sale) concluded that a leveraged buyout was possible.

A.R. Callow, Ltd
Harley-Davidson's early importer to South Africa, which made a slow start (39 machines in the first 18 months), though a visit from Alfred Child soon put things right. On the same trip, he sold 50 bikes to an African dictator.

Arena, Sam
A versatile rider who raced Harleys in hill climbs, TTs and flat tracks, winning races from 1934 until after 1950, when he took the classic racer's second career and became a dealer. According to author, Allan Girdler, Arena was one of the dealers who, after being shown the first XL-series Sportster, encouraged Harley to produce the stripped down XLC. That bike, with its tiny tank and (later) small

LEFT:
Harley-Davidson didn't completely ignore smaller bikes in the 1980s, though the Armstrong MT500 wasn't offered to the public. In fact, only the military qualified.

RIGHT: Art Deco styling used in the 1930s.

BELOW: The Arrow-Flite tank badge on a 1950s Duo Glide.

OPPOSITE: The atmospheric inlet valve, seen here on a 1905 single, severely limited revs and therefore power. However, a pushrod inlet valve soon replaced it.

headlamp, was the first with the classic Sportster look which survives today.

Armstrong, Erle
'Red' Armstrong was a racer in the pioneer days, riding Indians. He notably won the first race at the new board track at Tacoma, Washington, in August 1915. Average speed was 79.84mph (128.5km/h), a wheel length in front of Harley-mounted Otto Walker.

Armstrong MT500
A 30ci (500cc) off-road military motorcycle, designed by Armstrong Equipment of North Humberside, England. Harley-Davidson bought the design and manufacturing rights to it in October 1987, in an attempt to broaden its product base. The MT500 was powered by a single-cylinder Rotax engine.

Arrow-Flite Badge
Distinctive tank badge used by Harley which first appeared in 1953 and was used for the remainder of the 1950s.

Art Deco
For their first 30 years, Harley-Davidsons were hardly a byword in style. They were designed to be solid, reliable machines, and the standard colour (after the original grey) was olive green. But the 1930s heralded a new era of stylish design: more to the point, arch-rival Indian had been taken over by the du Pont chemical company, and was now offering a new range of bright colours. Rivalry between the two was so intense that Harley-Davidson had no choice but to respond, and from 1933 the side-valve Harleys came in a range of bold colours and eye-

catching graphics in the art deco style. This was a new era of speed and elegance, and Milwaukee did its best to keep up.

Artley, Roy
An early racer who was part of the famed Harley-Davidson Wrecking Crew racing team in 1920. In partnership with pioneer rider William J. Ruhle, he later took over the San Diego dealership of Hap Scherer, who himself had moved on to Los Angeles.

Artman, Bob
Appeared on a Harley in the 1933 Columbia movie, *State Trooper*.

Aspencade Motorcyclists Convention
(*See* Americade.)

Atmospheric Inlet Valve
Some early engines, including those of Harley-Davidson, relied on piston suction to open the inlet valve when, as the piston descended, the valve opened, with a light spring to close it again. It was a simple system, but limited engine speed to around 500rpm (that's 500rpm, not 5,000) which was fine for the pioneer single-cylinder machines. But as the demand for more power grew, so such lowly speeds became inadequate. This was made clear to William Harley in 1909, when his prototype V-twin produced little more power than the single from which it had derived. The answer, which appeared on the first production Harley twin in 1911, was a pushrod-operated inlet valve whose positive operation allowed higher engine speeds and thus more power.

Aurora Automatic Machine
Builder of the Thor motorcycle, the company also supplied rear hubs and

brakes to Harley-Davidson from 1910–13. William Ottaway was chief engineer with Aurora, but was soon poached by Bill Harley as an able assistant engineer. (*See* Ottaway, William.)

Aykroyd, Dan
Actor Dan Aykroyd, who starred with John Belushi in the film *The Blues Brothers*, rode a Harley. After John Belushi died, Aykroyd led the funeral procession on his bike.

Bagnall, William
President of the AMA from 1965, also editor and publisher of *Motorcyclist*.

Baker, Erwin
Indian racer in the second decade of the 20th century, who also made a speciality of endurance record attempts, which he completed on Ace, Excelsior, Henderson and Indian bikes. According to author Harry Sucher, he never did the same for Harley-Davidson because Walter Davidson wouldn't agree to pay his standard fees. Also co-designed (with Everett deLong) a 90-ci (1475-cc) V4 that was tried in an Indian Powerplus frame. DeLong later built a four-cylinder prototype for Harley-Davidson, which the Board rejected.

Balcom Trading Company
Established in 1945 to import Harleys into Japan, with headquarters in Takanawa, and 35 years later, it was still doing so. The founder was Richard Child, son of Alfred, who had built up Harley-Davidson business in Japan from nothing. (*See* Child, Alfred and Richard.)

Bang, Gary
Maker of aftermarket custom parts for Harleys.

Bangor Punta
A massive company that grew out of the railway industry, it attempted a hostile takeover of Harley-Davidson in 1968. Although Harley-Davidson needed cash at the time, Bangor Punta's approaches were unwelcome, and for good reason. Roger Hull, then editor of *Road Rider* magazine, considered that the reason was down to Bangor Punta's reputation: 'You can't blame the Harley-Davidson people for getting nervous,' he said later, 'Bangor Punta had a reputation for acquiring a company, squeezing it dry and then scrapping it for the salvage.' It was Harley-Davidson's fear of being sucked dry by Bangor Punta that drove it into the arms of AMF, which took over shortly afterwards. 'I decided to play the good guy,' said Rodney Gott, boss of AMF, 'and went to Milaukee to make friends with Harley management. I showed them how to combat hostile takeovers ... Bangor Punta was the bad guy and we were the good guys, so it was pretty easy to work out a deal.' The result was that Bangor Punta offered $23 million to take over Harley-Davidson, while AMF offered a million less, and still clinched the deal.

Bank of New England
One of the four banks that helped finance the management buyout of Harley-Davidson in 1981. Citicorp Industrial Credit was the main player.

LEFT: Sonny Barger, outside the Oakland Hell's Angels headquarters.

OPPOSITE: Bear Butte: a familiar landmark to anyone who has made the pilgrimage to Sturgis.

Barger, Sonny
Pictured below, the leader of the Hell's Angels in California during the 1960s. He rode a Harley. (*See* Hell's Angels.)

Barnes, Austin
A commercial photographer who, in 1938, was ordered by Walter Davidson to take pictures of a dismantled Crocker V-twin. Harley-Davidson was suspicious that the Crocker, which could outrun both Harleys and Indians at the time, was copying Harley-Davidson parts without permission. Austin's pictures showed nothing untoward.

Barr, Dave
Rode his 1972 Harley Super Glide 83,000 miles (133572km) around the world in the early 1990s. Barr had served in the US Marines and had lost both his legs below the knee while fighting in Angola. Determined to raise awareness of the plight of disabled people, he mounted his Super Glide (owned from new, and already with 150,000 miles/24140km under its wheels) and set off. Starting in Johannesburg, he rode up through Africa ('I'd never have made it through Zaire without belt drive'), where floods were the biggest problem, as there was no way of knowing how deep the water was. He was almost drowned when the bike keeled over and landed on top of him. Dave was just about able to keep his mouth above water level and hold on until a couple of villagers came along and rescued him. After Africa, Dave explored Europe and Scandanavia and crossed the Arctic circle before returning to the States. Then up to Alaska, back home to California, then down through South America, by boat to Hong Kong and by Harley through China, across the Gobi Desert and Mongolia, through Siberia, Moscow and Europe again. Dave Barr later wrote a book about his experiences on the road, entitled *Riding the Edge*, published by White Horse Press.

Barr, Frederick
Joined Harley-Davidson's engineering department in 1908, under George Nordberg, and was a key figure there for over 20 years. Also helped Nordberg and William Davidson to expand production in 1914. He finally left in 1929 to become chief engineer to the Sankyo company in Japan. His job there was to set up production of Harleys under licence at Shinagawa, a deal which had been masterminded by Alfred Child. Harleys were initially built up from parts imported direct from Milwaukee, but by 1935 Sankyo was making all-Japanese Harley-Davidsons.

Barry's Custom Cycles
Founded by Barry Matteson in Anchorage, Alaska in 1971, where he built and sold custom and tuned Harleys. He became a fully-fledged Harley-Davidson dealer in 1976, and despite Alaskan weather (not always conducive to motorcycling) did well.

Barth Manufacturing Company
Based in Milwaukee, where Bill Harley and Arthur Davidson both worked, met, became friends and went on to build a motorcycle.

Battistini
A top custom bike builder, based in the UK, which has set the standard for custom Harley building in Europe and beyond. Rikki Battistini now runs operations from the company's new headquarters in Glasgow, Scotland.

Battle of the Twins
Contemporary race series for twin-cylinder machines, dominated by Ducati in Europe and Harley-Davidson in North America.

Bauer, Louis
Boss of Indian in the 1920s, who in 1928 teamed up with Arthur Davidson and Schwinn of Excelsior to rescue the AMA.

Beals, Vaughn
A key figure in Harley-Davidson's turnaround in the 1980s, and manager in Milwaukee for AMF from 1975. Beals knew nothing about motorcycles (at least at first) but was a highly experienced and competent manager. One of his first actions was the so-called Pinehurst Meeting (*see also* below). Until then, Harley-Davidson had no long-term strategic plan, so Beals took all the senior managers up to Pinehurst, North Carolina for a week. At the intensive brain-storming session, it was agreed that Harley-Davidson urgently needed more modern, better-quality bikes to compete with the Japanese, and a ten-year product plan was drawn up. First, the existing V-twin was to be updated (which eventually saw light of day as the Evolution V-twin in 1983). Second, a whole range of all-new water-cooled engines would be developed, the NOVA programme. Far from being an AMF placeman, it was Beals who

originally proposed that Harley-Davidson senior management buy the company back from AMF, and it was he who negotiated the deal. As chairman of the newly-independent Harley-Davidson, he oversaw the huge changes in production through the 1980s which enabled the company to survive, putting quality first, with employee involvement and just-in-time methods.

Bear Butte Café
Part of the Sturgis experience, the intriguingly named Bear Butte Café sells a lot of beer while the South Dakota festival is on.

BEARS Racing
Acronym for the British, European and American Racing Series which allowed classic Triumphs, Ducatis and Harleys to battle it out on the track, without being constantly overtaken by Japanese bikes.

Beauchamp, Rex
Early 1970s flat-track racer.

Becker, Oscar
Oscar and Sherbie Becker were among Harley-Davidson's very first employees. An early photo of three of the Founders, plus eight workers grouped around a Harley sidecar outfit, shows Sherbie sitting on the bike with Bill Harley on the sidecar chassis.

Bedell, Alan
An endurance rider in the early days, in 1917 he rode 1,000 miles (1610km) non-stop at the Ascot Park race circuit, averaging 48.3mph (77.7km) for nearly 21 hours. He also set new records for the 24 hours (48.1mph/77.4km/h), 12 hours (49.5mph/80km/h) and 500 miles (50.1mph/81km/h).

Beer
There really was such a thing as Harley-Davidson branded beer and, given Milwaukee's importance in the American brewing industry, maybe it's not so surprising the Harley-Davidson version was not brewed in the Juneau Avenue plant alongside half-finished V-twin motors, but came in cans complete with the famous Harley-Davidson logo, now collectors' items (if they haven't already been drunk).

Bell & Howell Company
Harley-Davidson did subcontract work for Bell & Howell during the Second World War, producing telescopic equipment, among other products.

Belt Drive
For the first ten years, Harley-Davidson used the leather belt drive that was almost universal on early motorcycles; then, like almost everyone else, it went over to chain drive, which was more efficient, did not slip and could handle higher power. By the late 1970s, however, a popular conversion kit for Harleys was a toothed belt drive for both primary and final transmission which smoothed out the power pulses somewhat and needed minimal maintenance. The new FXB Sturgis of 1980 was the first production motorcycle to use a toothed belt (Kawasaki later used one on its GPz305). So successful was the new set-up that 20 years later every new Harley-Davidson comes with a belt-drive transmission.

Bendix Manufacturing Co
Harley-Davidson was forced to defend itself in a legal action instigated by Bendix in 1924 when Victor Bendix claimed that the Harley-Davidson clutch design infringed his own patent. Bill Harley had designed a multi-plate clutch very similar to the Bendix, and following legal advice, the Harley-Davidson version was rapidly redesigned and $18,000 was said to have been paid in damages.

Benton & Bowles
The advertising agency used by AMF, but which was not popular with Harley-Davidson. (*See* Advertising.)

Berndt, Ralph
Independent tuner who worked for Harley-Davidson but built his own race bikes. He was instrumental in the early career of talented racer Carroll Resweber in the 1950s. (*See* Resweber, Carroll.)

Bianchi, Alfredo
Designed the Aermacchi overhead-valve single-cylinder engine that later developed into the 15 and 21ci (250 and 350cc) singles sold by Harley-Davidson under its own name.

LEFT: Harley bicycles didn't last long, but the time may now be ripe for a Harley or Buell-branded mountain bike.

BELOW LEFT: There really *was* a Harley-Davidson beer.

BELOW: The earliest belt drive, and very different from Harley's later interpretation. There was no clutch: you just jumped on and got going.

OPPOSITE
TOP LEFT: The billet look, popular in modern customizing.

TOP RIGHT: A typical Big Twin (not a Sportster).

BELOW: Battle of the Twins racing enabled

Bicycles
For a short time, Harley-Davidson sold bicycles bearing its own badge, intending to cash in on the booming market for cycles and the brand's own rapidly growing status. According to the advertisements, it was 'the bike red-blooded fellows choose', and 'America's finest bicycle'. The idea was that red-blooded fellows would go on to buy Harley-Davidson motorcycles when the time was right, but the idea did not survive the tough times of 1920–21, and Harley withdrew from the bicycle business.

Big Twins
Generic term referring to all the larger V-twins in each era which included the 61/74ci (1000/1200cc) twins of the 1920s and '30s, the Knucklehead, Panhead and Shovelhead up to the early '80s and the Evolution and current Twin Cam twins. Sportsters are not Big Twins.

Billets
Parts machined from solid aluminium blocks, usually by CNC machines, form a large part of the aftermarket bolt-on custom parts market. Too many of these and your bike will be labelled a 'billet barge'.

Black Hills Motor Classic
The Black Hills rally began in 1938 but is better known as Sturgis, named after the small farming town in South Dakota which serves as a centre for the week's events. Now as famous as Daytona as one of the main (Harley-dominated) events in the American motorcycle calendar.

Blake, Robert
(See *Electra Glide in Blue*.)

Bleustein, Jeff
Harley-Davidson's head of engineering, an

associate professor of engineering at Yale, who was recruited in the middle of the AMF era by Ray Tritten. He really professionalized the engineering department at Milwaukee, bringing in highly qualified people who didn't necessarily know much about motorcycles, but had good track records in other fields of engineering. This wasn't always popular (the traditional Harley way was less academic), but in the end it bore fruit. Bleustein oversaw the development of the Evolution V-twin, which was crucial to Harley's 1980s turnaround, as well as the NOVA programme of water-cooled engines, which never got beyond the prototype stage.

Blue Knights
A motorcycle club whose members all belong or have belonged to the police, fire or emergency services.

BMW-Delco
The Second World War military motorcycle, built by the Delco company but closely modelled on the contemporary BMW flat twin, with shaft drive and telescopic front forks. The US Army was most impressed with this layout, and despite Bill Harley's view that an enclosed chain was just as practical as shaft drive, insisted that Harley-Davidson build something similar. This was the XA, which like the Delco was closely based on a BMW.

Board Track Racing
An early form of racing in America, using banked tracks made up of narrow wooden planks. This had a brief popularity, with tracks being built all over the country. But although smooth compared to contemporary roads, they offered far less grip, and the splinters could produce horrific injuries when riders fell off. They were soon superseded by oval dirt tracks.

ABOVE LEFT: The Blue Knights back-patch club.

ABOVE: The XA flat-twin engine unit, based on the equivalent BMW.

LEFT: An early board-track racer

OPPOSITE:
LEFT: Bootlegged (unlicensed) accessories were a major headache for Harley-Davidson, until it called the lawyers in.

RIGHT: The Boozefighters was one of the original post-war 'outlaw' bike clubs.

Boardwalk Custom Show
An integral part of Daytona Bike Week, the Boardwalk has grown from a side event into a show in its own right. Custom bikes of all shapes, sizes and types can be seen there every year, though it's fair to say that Harleys still predominate.

Boeing Aircraft Company
Boeing was just one of the companies for which Harley-Davidson did subcontract work during the Second World War, Allis-Chalmers and Massey-Harris being others. Shell parts, aircraft parts and over 13,000 truck winches were produced alongside 88,000 motorcycles.

Bond, John R.
Worked in Harley-Davidson's engineering department from 1946–47, and later edited *Road & Track* magazine. He noted the paradox that, although senior management encouraged new and innovative prototypes, these were never incorporated into production, hence the company's conservative model policy after the war.

Bonus Bucks
A dealer incentive scheme of the 1980s, except that the Harley-Davidson dealers didn't see it that way. Their profit margins on the big twins were cut by 5 per cent, but they could earn that back by taking part in company-run schemes: demo rides, advertising, stocking levels, service training courses. Feelings ran high, and 32 dealers actually broke away to form their own dealer organization, the Dealers Alliance. Relations between the DA and Harley-Davidson were strained until Harley-Davidson modified the Bonus Bucks scheme and made its official dealer organization elective. (*See* Dealers Alliance.)

Bootlegging
Harley-Davidson is one of the strongest brands in the world, so it's hardly surprising that many companies have tried to cash in on the name with unauthorized 'branded' products. Ranging from racist T-shirts to counterfeit spare parts and pot-smoking accessories, these reflected badly on Harley-Davidson itself. So from the early 1980s the company made a big effort to regain its own trademark, winning a big court case and getting into the licensed parts and accessories business. It meant strict control of what the Harley-Davidson logo could be appended to, such as lighters, pens, toys, T-shirts, the list is endless.

Boozefighters
One of the biker groups that grew up in the American West in the late 1940s/'50s, which (together with the events at Hollister) pre-empted mass media coverage of 'motorcycle hooligans'. Harley-Davidson was careful to distance itself from the 'outlaw' culture though, ironically, its success is now based on a stylized, sanitized version of exactly that.

Boston Consulting Group
Firm of industrial analysts which reported on the state of Harley-Davidson in 1978, towards the end of the AMF era. Its conclusion was that it could be viable, but only if AMF invested $60-70 million in the NOVA range of all-new water-cooled engines. Before NOVA bore fruit, AMF would have had to have undergone a 'valley of death' where investment was bound to exceed income. AMF, as we now know, chose not to go down that road.

Brinck, Eddie
A racer in the 1910s, Eddie Brinck was one of the original Harley-Davidson Wrecking Crew, which dominated US competition from 1915–21.

British Imports
America imported British motorcycles by the thousand in the 25 years after the Second World War. These were different from Harley-Davidsons, being light, nimble 350/500cc singles that introduced a new generation to bikes. By 1950, imports had captured 40 per cent of the market, which was why Harley-Davidson tried to have an extra import tariff imposed. Not until the late 1950s, with the XL Sportster, did the company have anything really comparable with a Norton or a Triumph. Maybe it was the experience of Indian, which had tried to meet the imports head-on with small home-built bikes, but which soon succumbed to bankruptcy, that encouraged Harley-Davidson to persevere with its conservative model policy.

British Motorcycle Dealers Association
Formed to represent those who sold British bikes in America after the Second World War. They had a legal battle with Harley-Davidson in 1951, when the company sought to have the duty on imported bikes increased from 8 to 40 per cent, and imports restricted to pre-war levels – a tiny fraction of what was being sold by then. The court found against Harley-Davidson, and in the next 20 years, dealers sold ever-increasing numbers of British bikes in the US.

Brough-Superior
The Brough-Superior motorcycle contrasted interestingly with Harley-Davidson. It was hand-built in England (allegedly by technicians wearing white gloves!) and it was expensive. By the late 1920s, most V-twins in Europe were sold as exotic, sporting machinery while in America big V-twins were the standard production motorcycle.

Buddy Seat
Before the buddy seat, the only way to carry a passenger was on a separate seat on the rear mudguard (uncomfortable in those hardtail days) or in a sidecar (rapidly losing favour in the early 1930s). The buddy seat was a joint invention of William S. Harley and Frank Trispel, foreman of the experimental department. It was an elongated saddle, flaring wider at the back, which was just big enough to take a passenger and rider in close proximity. Indeed, for many, this was part of the attraction, and a set of auxiliary springs could be clipped on to cope with the extra weight. It proved very popular, and was Harley-Davidson's best-selling accessory in the 1930s.

OPPOSITE

ABOVE: Brough-Superior typified the difference between European and American V-twins. On the smaller, twistier roads of Europe, V-twins were big, exotic and expensive. In America, they were a necessity.

BELOW: This 1937 ULH shows its gigantic white buddy seat to good effect.

RIGHT: The classic WR racer, still to be seen on the track today.

Burkey, Eric
President of the IAM Lodge at the York factory. With others at Harley-Davidson, he went on a tour of the Honda factory at Marysville, Ohio to take a look at Japanese production methods. He later recalled his amazement that Harley-Davidson engineers from York were making hurried sketches (no cameras allowed) of how the Honda workers were bleeding brakes. This was the same method that had been used at Milwaukee for years, but no one at York appeared to have heard of it. This anecdote is a good illustration of one of Harley-Davidson's central problems at the time – two plants hundreds of miles from each other which didn't appear to be on speaking terms.

Buschmann, John
Worked for Harley-Davidson for just three months from late 1965–66. He was behind a stolen parts scam in which new parts were smuggled out of the factory and sold to the unsuspecting. The alarm was raised when one Milwaukee dealer was offered 10,000 Harley spark plugs at 50 per cent of the cost of production. It transpired that 29,000 plugs were missing. The FBI was called in, thousands of dollars worth of missing parts were traced, and Buschmann went to prison.

Byrnes, Ed
Ed 'Kookie' Byrnes rode a Harley-Davidson Topper scooter in the TV series *77 Sunset Strip*. In spite of this, it still wasn't a great success.

Cabbage Patch
Infamous Daytona campsite called Sopotnik's Cabbage Patch, where the main attraction during Bike Week is the Women's Coleslaw Wrestling (it is a cabbage farm after all!), where clothes are optional.

Cable, Charles
Charles Cable ran an import-export business from San Francisco in the 1920s, and part of his business was the export of Harleys to Mongolia. The factory was puzzled as to why he never ordered any spare parts, and it turned out that he was diverting machines to Japan instead. In the light of this, and because of the lacklustre performance of the current Japanese importer, Alfred Child was put in charge of a new Japanese operation, beginning an era of healthy Harley-Davidson exports to the Far East. (*See* Child, Alfred.)

California Bobber
Arguably the first real custom bike, or it at least heralded the birth of the modern custom movement. It first appeared in California in the late 1940s, and the look rapidly spread across America and then the rest of the world. The easiest way to make a big Harley or Indian lighter and faster was to dump all the superfluous bodywork, so front mudguards, screens and panniers were all removed. The rear mudguard was 'bobbed' or 'chopped' (hence 'chopper') down to the bare minimum, in keeping with the stripped-down look. As years

went by, the chopper or bobber look became ever more stylized and elaborate, popularized by Peter Fonda's Captain America chopper in the film *Easy Rider*: but the California Bobber was where it all started.

California Highway Patrol
CHiP has long been a high-profile user of Harley-Davidsons, and though the company lost the contract to Kawasaki in 1986, it has since been won back.

Cars
The advent of cheap mass-produced automobiles spelt death to many motorcycle makers. By the early 1920s there were around 400 makes of car on sale in America, and you could buy a good one for $400, the price of a sidecar outfit. Through drastic cost-cutting, Harley-Davidson managed to survive, but it was a close call, and the market had changed for good. In North America, more than in Europe, the days of the motorcycle as basic transport were over; they were increasingly becoming sporting machines for enthusiasts, and this was reflected in Harley-Davidson's concentration on big bikes.

OPPOSITE: Paintwork apart, this California Bobber demonstrates how customizing started. Note the lack of front mudguard, the 'chopped' or 'bobbed' rear guard, high bars and fat tyres. However, fancy colour schemes came later.

RIGHT: CHiP – the California Highway Patrol – was a long-term prestigious user of Harleys, until it changed to Kawasakis. After some hard work, they are back with Harley again.

Celebrity Owners
Countless celebrities are or have been Harley-Davidson owners, and there are now more than ever before. For some, it's a piece of strategic image enhancement, for others, a genuine love of the bike. Whatever the reason, it moved one rider to observe that Harleys in Hollywood are now like a certain part of the human anatomy – everyone's got one!

Music

Bryan Adams
Aerosmith
Jon Bon Jovi
Cher
Eric Clapton
David Crosby
Neil Diamond
Fabulous Thunderbirds
The Grateful Dead
Isaac Hayes
Peter Hook (Monaco)
Billy Idol
Michael Jackson
Billy Joel
Ronan Keating (Boyzone)
Tommy Lee
John Cougar Mellencamp
George Michael
Larry Mullen (U2)
Willie Nelson
Olivia Newton-John
Robert Plant
Lou Reed
Steppenwolf
Barbra Streisand
Dave Stewart (The Eurythmics)
The Stray Cats
Terence Trent D'Arby
Tonya Tucker
Eddie Van Halen
Dave Vanian (The Damned)
Stevie Ray Vaughan
Paul Young

Actors

Dan Aykroyd
Ann-Margret
Antonio Banderas
James Caan
Martin Collins
Clint Eastwood
Peter Fonda
Harrison Ford
Goldie Hawn
Don Johnson
Jay Leno
Ewan McGregor
Liam Neeson
Priscilla Presley
Micky Rourke
Kurt Russell
Arnold Schwarzenegger
Steven Segal
John Gordon Sinclair
Sylvester Stallone
Liz Taylor
Richard Thorp
Bruce Willis

Sportsmen

Eric Cantona
Chris Eubank
Paul Gascoigne
Ian Wright

Motor Racing

Johnny Herbert
Damon Hill
Eddie Irvine
Adrian Reynard
Mika Salo

Others

David Copperfield (magician)
Michael Flatley (Riverdance)
Richard Young (paparazzo photographer)

Ex-Harley Owners

Sir Malcolm Campbell
Malcolm Forbes
Clark Gable
Michael Hutchence
Charles Lindbergh
Elvis Presley
Roy Rogers

Chann, Jimmy
A successful rider for Harley-Davidson after the Second World War, winning the 1947, '48, and '49 Grand National Championships.

Child, Alfred
Alfred Child was a supersalesman who did much to spearhead Harley-Davidson's exports in the 1920s and '30s. Born in England, he worked his passage across the Atlantic at the age of 16 and after working as a butler, handyman and seller of cycle parts, joined Harley-Davidson. After acting as Arthur Davidson's assistant, he was sent to Africa, which he explored riding a J-model and sidecar. Starting in Cape Town and finishing in Cairo, he took orders for 400 bikes and galvanized many dealers into action. Then Davidson sent him to Japan, where Child was to remain until war broke out, first importing Harleys, then setting up a production facility for which drawings and machine tools were bought from Milwaukee, and the Sankyo company began to make all-Japanese two- and three-wheelers based on Harley-Davidson designs. After a disagreement, the contract was terminated and Alfred Child was again Harley-Davidson's Japanese importer; but after the Second World War, Child turned up in the USA as the importer for BSA! It was his evidence (citing his knowledge of Harley-Davidson's restrictive practices) that helped persuade the Tariff Commission to turn down Harley-Davidson's request for higher tariffs on imported bikes in 1951. Alfred Child lived to a ripe old age into the late 1970s. He had had an eventful life, to say the least.

Child, Richard
Alfred Child's son, he worked closely with his father importing Harley-Davidsons into Japan. His 400-mile (645-km) test ride of one of the first 61E Knuckleheads to arrive was not a success: it leaked oil, broke valve springs and needed frequent tappet adjustments. Despite this experience, Richard Child went on to become the official Harley-Davidson importer for Japan after the Second World War.

Chopper
A development of the original California Bobber (*see above*). Through the 1950s, '60s and '70s, the chopper grew ever more elaborate, with the movement spreading right across the world. Bars grew higher, as did seat backs, and multi-coloured metalflake paint jobs and intricate chrome work or engravings appeared. For many, the definitive chopper, with its high bars, tombstone seat and feet-forward stance, is the Panhead Harley ridden by Peter Fonda in the film *Easy Rider*.

OPPOSITE: It may seem amusing to us now, but this was regarded as the ultimate custom machine in the mid-1970s. Everything, apart from the brakeless front wheel, is to excess – forks, lights, pipes and sissy bar. It was at this point in time that the question was asked: are these motorcycles works of art or not?

Church, Gene

Dominated the Battle of the Twins series (*see above*) in the mid-1980s, winning in 1984, '85 and '86 on the Harley XR1000-powered Lucifer's Hammer. A former dirt-track rider, Church went on to race in Grand Prix.

Class C Racing

This was a racing class thought up in the early 1930s by Harley-Davidson and Indian for production machines, as opposed to expensive competition specials. The idea was to reduce the cost of racing, as both factories were finding the cost of special racers prohibitive. Instead, standard production 45ci (750cc) side-valve and 30ci (500cc) overhead-valve bikes were eligible for flat-track racing. Class C TT events (what we would now call motocross) and hill climbs also allowed 61, 74 and 80ci (1000, 1200 and 1300cc) twins to compete in Class C, which had its first season in 1933. The rules, of course, were explicitly designed to suit Harley-Davidson and Indian twins.

Colours

Harley-Davidson first offered a wider range of colours in 1933, in direct response to Indian's initiative a year or two earlier. Traditionally, Harleys came in drab grey or green, but for 1933 the twins could be had in silver and turquoise with black and gold pinstriping; black and mandarin red with gold striping; sunshine blue and white with gold striping; and police blue and white with gold striping. There was no extra charge for any of these. It was also the year of the handsome art deco bird motif on the tank. In fact, Harleys had never looked better.

Constantine, Arthur

An assistant engineer at Harley-Davidson from 1923, he was a graduate engineer fresh from Buick. He resigned the following year, after Bill Harley and Arthur Davidson rejected a 45ci (750cc) side-valve prototype he had built. Constantine went to Excelsior, where he produced the successful Super X. Harley-Davidson's own 45ci twin was brought out in direct response.

Crewe, Della

Completed a 5,400-mile (8690-km) journey across the US on a Harley-Davidson sidecar outfit in 1915, with only her dog Trouble for company. Della Crewe had had only ten days' riding experience before she left Waco, Texas on her 1914 outfit. She had considered taking the train for her cross-country tour, but rejected it as 'uninteresting'. With only dirt roads to follow, little more than rutted quagmires due to a very wet winter and spring, progress was painfully slow, particularly on the first leg of the journey to Dodge City. At one point, Della had to abandon the road and head off across the wheat fields for 4 miles before finding a useable track. The weather and roads were better as they headed north through Missouri, then crossed the Mississippi and continued on to Chicago and the Harley-Davidson factory in Milwaukee. In Indiana, Trouble was threatened with quarantine due to contact with foot-and-mouth disease, and by the time the pair reached Ohio it was November and the weather was closing in with drifting snow sending them frequently into ditches. Finally, Della and Trouble made it to New York after 5,378 miles (8655km) and no major breakdowns. 'I had a glorious trip,' said Della to reporters, 'I am in perfect health and my desire is stronger than ever to keep going.'

Crocker

A high performance V-twin built by Albert Crocker and Paul Bigsby in the late 1930s. Much to Harley-Davidson's chagrin, it was

The Crocker was a thorn in Harley-Davidson's side, being faster than anything out of Milwaukee, thanks to its highly tuned overhead-valve V-twin motor. Walter Davidson did his best to get production stopped.

much faster than any tuned Harley (or Indian for that matter), being capable of a genuine 100mph (161km/h). The Crocker was an expensive limited-production machine, but Walter Davidson was so incensed at its performance that he tried to get production stopped. He threatened Budd and Kelsey-Hayes (who supplied wheel rims to Crocker) and Linkert (carburettors) that if they didn't stop doing so their lucrative Harley-Davidson supply contracts would cease. They agreed, though the Crocker stayed in production for a little while longer.

Curtiss, Glenn
A pioneer aviator who rode a Harley.

Custom Bikes
Harley-Davidson has long been central to the custom bike movement, particularly in the USA. It grew out of the original California Bobber (*see above*), into the chopper, and later into countless variations such as street fighters and retro bikes. In America, at least, the Harley is still the bike of choice for custom builders.

Cycle Tow
Predecessor of the Servicar (*see* below), the Cycle Tow was basically a standard motorcycle with two outrigger wheels fitted to the rear of the frame. The idea behind this four-wheel motorcycle was that mechanics could ride it out to customers' cars as a two-wheeler, then hinge down the outrigger wheels, hitch the whole device to the car's rear bumper and drive back to the garage, cycle in tow. Once the car had been serviced, it would be driven back to the customer and the whole operation would be reversed. Not many Cycle Tows were sold, but the idea soon led to the immensely successful Servicar, which became a virtual American institution, remaining in production into the 1970s.

Davidson Family
Hailing from Aberdeen, Scotland, the Davidsons emigrated to America and settled in Milwaukee in 1871 or '72. William Davidson and his wife had two daughters and three sons, Arthur, Walter and William A. In the fledgling Harley-Davidson company, Arthur was the salesman, Walter the general manager (he built the first bikes) and William A. the works manager. Walter and his wife, Emma, had three sons, Gordon, Walter Jnr and Robert.

Davidson, Arthur
Co-founder of Harley-Davidson with Bill Harley, Arthur Davidson was a natural salesman who, in the words of author Herbert Wagner was 'an out-front brassy little guy. A handshaker. He loved people ... He liked a good joke and he loved telling stories.' It was his extravert personality that meshed so well with the engineering skills of his friend, Bill Harley, and played a key role in the early success of the company. While his brothers were building bikes and Bill was developing them, Arthur was out on the road recruiting reliable dealers and promoting the company through advertising. He recognized the importance of training, and founded the Harley-Davidson service school. Like Bill Harley, Arthur loved the

outdoor life. He died in December 1950, the victim of a road accident.

Davidson, Bill
Son of Willie G, and the fourth William Davidson to work for the company, he joined the 'family firm' (which of course it no longer was) in 1984 to oversee the national demo ride programme and gain experience in the advertising and promotion departments. In mid-1991 he was nominated head of the Harley Owners Group (HOG).

Davidson, Gordon
The son of Walter, he joined the sales department in 1930 after graduating from the University of Pennsylvania. Seven years later he was a company director and was appointed vice-president of manufacturing in 1942. More at home on the shopfloor than in the office, he died in 1967. His brother, Walter C., was much like their father, a natural salesman fond of company and the odd drink. Young Walter rose to become vice-president and sales manager. He died in 1974. Less is known of their cousin Allan, William A.'s son. He too worked for the family firm, but sadly died young.

ABOVE: Willie G. has to be the most famous living Davidson, being the grandson of the original William A. Davidson. Not only is he the stylist behind countless bikes of the 1970s and '80s (notably the Super Glide and Low Rider), but he is a hard-working ambassador for the company, attending rallies and meeting the customers.

RIGHT: Walter Davidson, second of the Davidson brothers, pictured after winning an endurance event on an early single.

Davidson, Janet
Aunt to the Davidson brothers, her connection with the Harley-Davidson story is that she painted the red pinstriping and Harley-Davidson logo onto the tank of the very first 1903 machine.

Davidson, John A.
Son of William H., he was appointed president of the company in the late 1970s, after a succession of non-family presidents failed to solve the company's problems. He resigned just before the management buyout was completed in June 1981.

Davidson, Karen
Daughter of Willie G., who trained in fashion design before joining the Harley-Davidson clothing department and becoming involved in Harley-Davidson's own MotorClothes range.

Davidson, Walter
Second of the Davidson brothers and president of the company, Walter learned his trade of machinist in the railway workshops of Milwaukee. It was he who built up the first motorcycle that his younger brother, Arthur, and William Harley designed. Often described as the most reserved of the four Founders, he turned out to be a highly competent rider, and did well in early endurance events. Walter was still president of Harley-Davidson when he died in February 1942.

Davidson, William A.
Eldest of the Davidson brothers and company works manager and vice-president, William was already foreman of a railway workshop when his younger brother and Bill Harley were building their first motorcycle. Consequently he brought a wealth of practical experience in production, metallurgy and management to the fledgling company. Nicknamed 'Big Bill' or 'Old Bill', to differentiate him from his son, William H., was a big man who loved outdoor pursuits such as hunting and fishing. Veteran employees remember him as someone who always remembered them. He died in 1937.

Davidson, William H.
Son of William A. Davidson, he succeeded Walter Davidson as president in 1942. He was still in the job at the time of the AMF deal, and urged shareholders to accept the takeover. He retired shortly afterwards, in 1971.

Davidson, Willie G.
Son of William H., and the man behind many key bikes in Harley's recent history, Willie G. Davidson styled the original Super Glide (1970) and Low Rider (1977), and remains head of the styling department. He trained in design, and worked for Ford and Brooks Stevens before joining the family firm. Harley's survival through the 1970s and revival in the '80s owes a great deal to Willie G., whose stylistic skills have allowed the company to produce apparently new bikes using many existing (sometimes outdated) parts. Just as important, he has remained in touch with what riders want – his Super Glide of 1970 was the first factory interpretation of what customizers were already doing. In the process, he has managed to transform himself from an earnest-looking young man with horn-rimmed spectacles and a sensible haircut into a grey-bearded biker. That in itself puts him in tune with the typical 21st-century rider of a new Harley, i.e. an affluent middle-aged professional.

Davis, Bill
Davis is a customizer who could fairly claim to be the godfather of the current line of Softail Harleys. In the early 1980s, he built a bike with the rear suspension units hidden horizontally under the gearbox, while the swinging arm and frame were reworked to give the impression of a hardtail rear end, but with more contemporary comfort. The story goes that Vaughn Beals saw this bike at a rally and was so impressed that he ordered a factory version. The official Softail appeared in 1984, proved a tremendous hit, and has been a big part of the Harley-Davidson line-up ever since.

ABOVE: The Softail was originally designed by a customizer named Bill Davis. Vaughn Beals spotted it, and the rest is history.

OPPOSITE: The reality of Daytona Beach, somewhat different from the free cruisin' image usually depicted.

Daytona Bike Week
What started as beach races at Daytona, Florida progressed to the purpose-built Daytona Speedway track and the famous Daytona 200 race. But a major festival of motorcycling has developed around the racing which, although it has seen increasing numbers of Japanese bikes in recent years, is still really Harley-Davidson-dominated. Apart from cruising the flat sandy beach in front of the town (recently restricted), there are custom shows (the Rat's Hole and Boardwalk), innumerable trade stands and Main Street, which for Bike Week is solid with Harleys.

Dealer Advisory Council
The company-sponsored association for US dealers. Until the mid-1980s, its members were appointed by the company, but they were elected by the dealers after their revolt of 1983. It still represents the vast majority of North American Harley-Davidson dealers.

Dealers Alliance
The national Harley-Davidson Dealers Alliance, formed in 1983 by 32 American Harley-Davidson dealers in protest to the new Bonus Bucks incentive. Three hundred had joined within two years, 81 per cent of whom declared they had no confidence in Harley-Davidson management; but over the years, relations did mend. The company's Dealer Advisory Council became more democratic and communication between the company and its US dealers improved. (*See* Bonus Bucks.)

Dean Witter Reynolds
Firm of investment advisors that organized Harley-Davidson's refinancing package in early 1985, a key figure being Steve Deli of the firm's Chicago office, who was able to bring together Harley-Davidson and Heller Financial, another investment bank that was interested in putting money into the company. The original financier, Citicorp, pulled out in 1984, and finding another bank was the only way the company could be saved.

Deathshead
A winged skull which appears on the Hell's Angels' backpatch and is often painted elaborately on their gas tanks. It is patented worldwide and woe betide anyone copying the design or wearing anything similar on their jackets.

De Savary, Peter
A millionaire property developer and Harley aficionado who opened up his Civil War mansion in Berkshire, England for a number of Harley rallies in the late 1980s and early 1990s. He has an impressive collection of vintage automobiles and motorcycles.

A Harley accessory? ABOVE is British entrepreneur Peter de Savary, sporting a neatly trimmed beard, while RIGHT is American customizer Rick Doss, with a rather fuller version.

OPPOSITE: The cockpit of the FLT, with all the accessories you would need for crossing the American continent. What else can you spot besides the extra mirrors and mobile phone?

Dixon, Freddy
English rider and tuner of Harley-Davidsons in the 1920s, he scored many race wins, both solo and sidecar-mounted, notably at Brooklands. He also modified an early eight-valve Harley racer to take twin carburettors.

DKW
German motorcycle maker, established in 1919, whose pre-war RT125 two-stroke was awarded to various Allied manufacturers as war reparations. BSA produced it as the Bantam, the Russians made their own version and for Harley-Davidson it became the Model S, or Hummer.

Doss, Rick
Famed American East-Coast builder and designer of custom parts under the shop name of 'Rick's Custom Harlees' (sic). He is also a consultant to the multimillion-dollar customs parts company, Custom Chrome Inc.

Dressers
Dressers are Harley-Davidsons which have been literally 'dressed' for touring with screen, panniers, spotlamps, large comfortable seat and maybe a fairing. The origins lie in 1920s and '30s America, when riders began to add practical touring accessories to cope with the long distances in that big country. It developed a look all its own, the antithesis of the stripped-down bobbers and choppers after the Second World War. But the dresser survived to become an integral part of the world of Harley-Davidson: the Electra Glide, Road Glide and Road King all draw on this tradition, being ready-made dressers (in various states of undress) available direct from the factory.

Harley-Davidson
USA

Their credo is violence...Their
is hate... and they call themsel
THE WILD ANG
THEIR GOD IS
AND—LIKE RABID
...LUST IS THE LAW
THEY LIVE BY!
ROGER CORMAN
AN AMERICAN INTERNATIONAL
DEVILS ANGELS
IT'S THE
Snarling Engine
NOW THE ARM
THE OUTLAWS
IN THE MOST
ACTION PICTU
OF ALL TIME!

Easy Rider
The film in which Peter Fonda and Dennis Hopper ride choppers across America under big blue skies, 'looking for adventure' as the Steppenwolf soundtrack goes, also featured one of the most famous Harleys of all time – Captain America. The appearance of this definitive chopper, with its high bars, raked forks and easy riding stance, made a huge impression on riders, not just in America but all over the world. The film combined late 1960s hedonism with a traditional, almost Wild West theme of two drifters in search of freedom. The only difference was, instead of horses, they rode Harley-Davidsons.

Ekins, Bud
Famous desert competition rider on Harley-Davidsons and friend of the late Steve McQueen. Ekins' collection of Harleys has often been seen in films, notably the bike ridden by James Stewart in *The Spirit of St Louis* (the Charles Lindbergh story) and the American TV series *Nichols*, in which James Garner rides a Harley-Davidson single.

Electra Glide in Blue
Another world-famous film starring a Harley. In 1971, Robert Blake played a Harley-mounted motorcycle patrolman who dreams of his perfect bike, an Electra Glide in blue. The bike's role in the film is to point up Blake's own dilemma, the choice between law enforcement and the pursuit of the road to freedom.

Electric Lighting
This arrived in 1915, along with a three-speed gearbox, to keep up with rival manufacturers and replace the soon to be outmoded acetylene lights.

Electric Starter
This arrived in 1965 on the aptly named Electra Glide. Harley-Davidson could not be accused of being late with electric start, which was only just appearing on Japanese machines at the time. It needed to be beefy to turn over the 74ci (1200cc) V-twin, and the big starter and new 12-volt electrical system together added 75lb (34kg) to the weight of the machine.

Engines
F-head Singles & Twins: 1903–1929
The name given to the first engines used by Harley-Davidson, so-called because of

OPPOSITE: Captain America: Peter Fonda's bike from *Easy Rider* inspired a whole generation of custom builders, both outside as well as within the USA.

Electric lighting on a 1925 J-model (LEFT) and electric starting on a 1965 Electra Glide (ABOVE). What's 40 years when the company has lasted for nearly 100?

the 'F-head' layout, with side-mounted exhaust valve and pushrod-operated overhead inlet valve, though very early motors had a vacuum-operated inlet valve.

Side-Valves: 1927–1948

Side-valve singles and V-twins were the basis of Harley-Davidson's engineering policy from the late 1920s to the late 1930s. A major consideration was that arch-rival Indian used side-valve engines, and they were often faster than the equivalent Harley.

Knucklehead: 1936–1947

Name given to Harley-Davidson's overhead-valve V-twin, introduced in 1936, because the rocker covers were thought to resemble a pair of knuckles. It was an engineering milestone for the company, being faster and more advanced than any of the opposition (i.e. Indian) and played a great role in transforming Harleys from reputedly being something of an old man's bike to a real sportster. It was replaced by the Panhead in 1947. (*See also* Big Twins.)

Panhead: 1947–1966

Aluminium cylinder heads for improved cooling and hydraulic tappets were added to the existing bottom end. Also cooking-pan-shaped rocker covers, hence the nickname. A clean-looking motor, but not without its teething troubles.

Shovelhead: 1966–1983

The nickname for the updated Big Twin of 1966, which adopted a Sportster-style top end to boost power. Power was increased again, to 65bhp in 1968, though capacity remained at the now familiar 74ci/1200cc. It was a long-lived engine, surviving right throught to 1983.

Evolution: 1983–1998

Substantially an updated 82ci (1340cc) big twin, which formed the bedrock of the company's turnaround in the 1980s. It was originally intended as a stop-gap before a range of all-new liquid-cooled engines were unveiled, but stayed in production as Harley's sole engine for 15 years. With aluminium-alloy cylinder barrels, a lighter

OPPOSITE
FROM TOP LEFT to RIGHT: F-Head: The first twin with overhead inlet, side exhaust valves. Knucklehead: The first ohv twin and a massive leap forward. Panhead: Hydraulic tappets introduced not only automotive technology but also teething troubles. Shovelhead: More power to cope with more weight. BELOW: The Shovelhead came with aluminium heads. The SU carburettor is non-standard.

ABOVE: The Evolution motor heralded a whole new era of reliable, oil-tight Harleys. It actually has much in common with the Shovelhead it replaced, but the initial quality was higher, as were the engineering tolerances.

ABOVE RIGHT: The Twin Cam is the current Big Twin, the largest yet at 1450cc (88ci). Like every other engine apart from the Knucklehead, it's more of an evolution than a revolution, though the 88B balance-shaft version (which appeared in 1999) is something of an innovation for Harley.

valve train and closer tolerances, it was more powerful, more oil-tight, better built and more reliable than its predecessors. Harley-Davidson owes a great deal to this engine.

Twin Cam 88: 1998–
Harley-Davidson's current Big Twin, still an air-cooled 45-degree V-twin with hydraulic tappets, but really an all-new engine compared to the Evolution, which it replaced over 1998/99. The biggest Harley yet, at 88ci (1450cc), with a bigger bore/shorter stroke, twin camshafts and a choice of carburettors or fuel injection. A balance shaft version, the 88B, was launched in 2000, though at the time of writing, the 883/1200 Sportsters are sticking with their Evolution engines.

Engraving
A customizing technique, using traditional engraving methods to decorate bare metal, usually engine crankcases and exhausts. It was pioneered by English customizer John Reed, who persuaded gun engraver Don Bloxidge to try his hand at bikes. The technique reached its apogee in the 1970s and '80s, but since then the use of heat-resistant powder coating has made possible alternative ways of decorating engine cases.

The Enthusiast
Harley's own magazine, launched in 1916 and sent out to all Harley-Davidson dealers and registered owners. Although not always a regular publication, and it could hardly be called non-partisan, it carried on for decades afterwards.

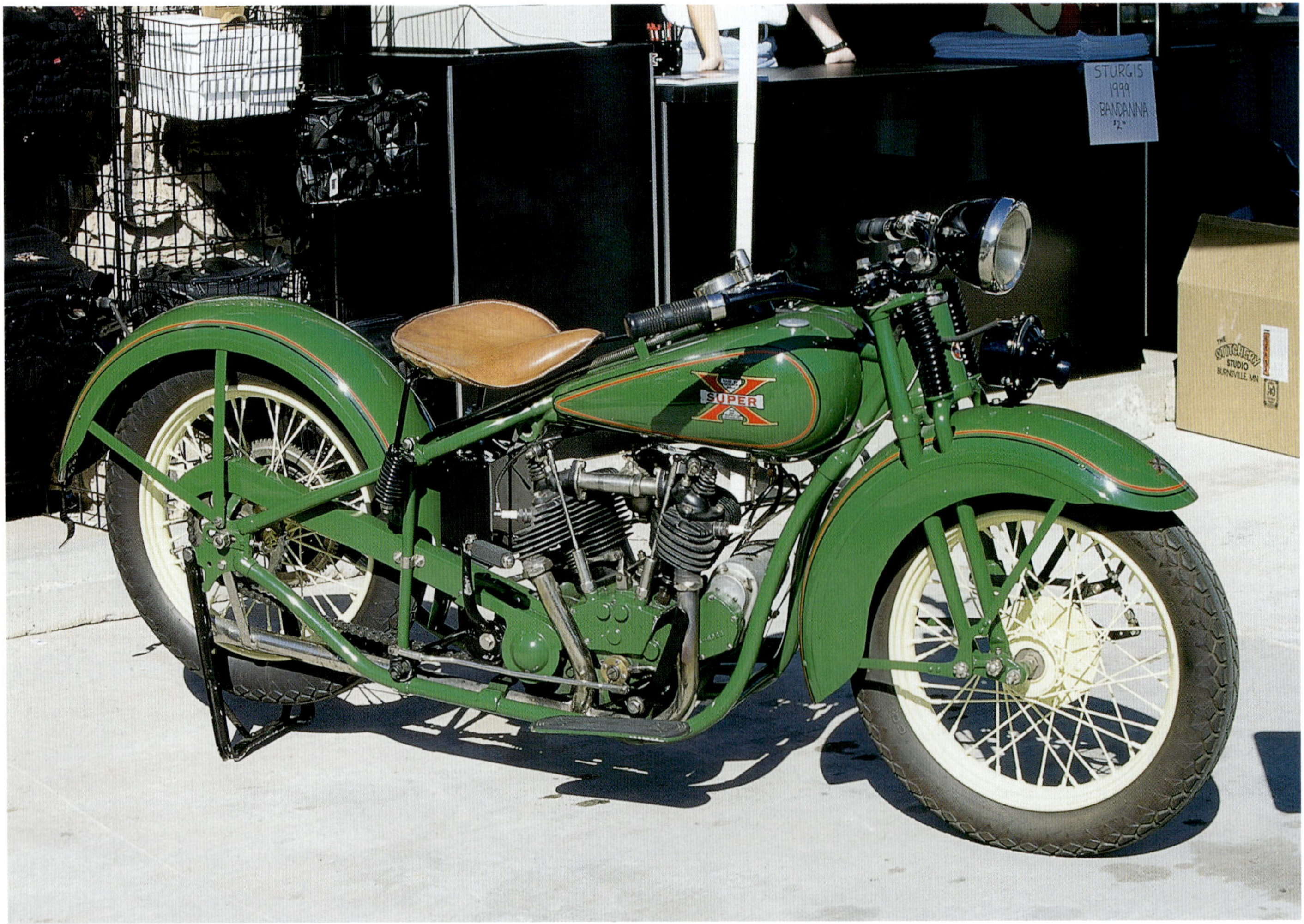
STURGIS
1999
BANDANNA
SUPER
STUDIO
BURNSVILLE, MN

Excelsior
A key early competitor of Harley-Davidson. Excelsior was formed in 1908 and later taken over by Schwinn bicycles. It produced a full range of machines but failed to survive the Depression, and was closed down by Schwinn in 1931.

Excelsior-Henderson
An American-made competitor to the Harley in the late 1990s. Harley-Davidson's turnaround in the 1980s and later success attracted some entrepreneurs into investing in a home-built competitor, notably the Polaris company (which makes the Victory) and the Hanlon brothers, who revived the Excelsior-Henderson name around a brand new 84ci (1386cc) V-twin. Production began in 1998, but at the time of writing is suspended while Excelsior-Henderson is in financial trouble. The future remains uncertain.

Factory Customs
These are factory-built Harleys (or indeed any bikes) which follow current customizing trends. For Harley-Davidson, this started in 1970 with the Super Glide, whose long low look reflected what customizers of the time were doing. It was the first time Milwaukee recognized the custom movement, despite the fact that through the 1960s there were probably more Harleys being chopped than any other bike. They even beat the Japanese on this one, who later began to produce their own 'factory customs', which were really standard bikes with higher bars and a dash of extra chrome. It was also the first major project to emerge from the styling department headed by Willie G. Davidson. The Super Glide wasn't actually a huge

OPPOSITE: The Excelsior Super X out-performed equivalent Harleys.

ABOVE LEFT: A modern Excelsior-Henderson of 1998. Sadly, the task of competing on Harley's home ground proved more difficult than was expected.

ABOVE: The lavishly painted tank on this 1976 bicentennial bike is a wonder to behold.

success, but it did pave the way for a whole stream of factory customs from Harley-Davidson, notably the low-seated Low Rider of 1977, the Wide Glide of 1980 and the Softail of 1986. Each one followed a particular custom look, and all were successful, so much so that factory customs now play a major part in Harley-Davidson's model range. It is hard to believe now that in 1980 Harley-Davidson management was reluctant to allow the Wide Glide into production, as its wide-spaced raked forks were thought to be too strongly connected to the outlaw element of biking.

Fat Bob

A renowned Harley dealer, living in Dorset, England, and famous for his keen interest in blueberry muffins and thick shakes. In reality, a Fat Bob is the common wide style of fuel tank with two filler caps, one each side. One cap may be a fake, the style harking back to the days when gas was held in one side, oil in the other. The style is still used on Harley-Davidsons today, the left-hand (fake) cap being used to house a fuel gauge.

ABOVE: One of Arlin Fatland's custom Harleys. Note the magnificent spoked wheels.

LEFT: The Fat Bob tank harks back to the days when fuel tanks held oil as well as gas.

OPPOSITE: A familiar decorative theme on custom bikes, especially in the 1970s. This bike's so hot it's on fire!

Fatland, Arlin

Famous custom bike builder from Denver, Colorado who has been operating from his shop, Twowheelers, for almost 30 years. He still turns up at Sturgis with head-turning custom Harleys.

Flames

The flame device is the most traditional of all custom paint treatments. It harks back to the paint jobs found on hot-rod cars in California in the 1930s and 1940s.

Focus Flow Processing

Part of Harley-Davidson's transformation of its manufacturing operation in the 1980s. It had traditionally built components in batches, which had to be moved all over the plant for various operations. This meant many parts were either queueing up at production bottlenecks, or held in stock until they were needed, which wasted time and money. Reorganizing production into a continuous flow, with each employee responsible for several operations, meant different parts could be produced as needed, allowing the company to respond more efficiently and speedily to orders. Production costs and times were cut, inventories slashed, and quality improved.

Forbes, Malcolm

Millionaire Malcolm Forbes didn't start

riding until late in life, but made up for it with a Harley stationed at each of his houses, not to mention one on his private plane and another two on his yacht. He founded a club, the Capitalist Tools MC (which survives him) and had a hot-air balloon made in the shape of a Softail.

Fordyce, Skip
A popular and successful pioneer Harley-Davidson dealer in Riverside, California. He caused a scandal after the Second World War by taking on Triumph machines as well, and a horrified Harley-Davidson management threatened to withdraw his franchise. Fordyce pointed out that Harley had nothing to offer new riders in the increasingly popular 30–45-ci (500–750-cc) market, adding that he could take them to court for restraint of trade. Harley-Davidson withdrew its objection.

Forecar
Introduced in 1913 as an alternative to the conventional sidecar outfit. It replaced the motorcycle's front wheel and forks with a large load-carrying box and two steerable wheels connected by an axle. Everything aft of that was pure standard motorcycle. The Servicar which superseded it was simply the same idea back to front. (*See* Servicar.)

Forty-Five OHV (Prototype)
Although the standard long-lived Forty-Five, and even its successor the K-series, were all side-valves, there was talk of an overhead-valve version as early as 1937. Bill Harley suggested that they should use the standard bottom end but with new cylinders and heads. It would use the standard frame, in strengthened form, and produce a projected 30hp even in a mild state of tune. Bill Harley pushed hard for this model, pointing out that although it would cost about the same to make as the 74ci (1200cc) side-valve, it would also produce as much power and would be more reliable into the bargain (there were continuing piston problems with the big twins). In August, a preliminary prototype, which had been built based on a contemporary 45 ohv hill climber, with exposed valvegear but giving a good idea of how a 45 ohv might perform, gave 29hp.

However, Arthur Davidson disagreed, maintaining they should concentrate on the existing bikes and use the money saved to produce a new smaller lightweight machine. Bill countered that an all-new design would take two years to come to fruition and cost more than his idea anyway. But he was evidently on a losing streak. Walter Davidson voiced doubts that the existing 45 gearbox would be able to accept the ohv's extra power, for even in the modestly powered Forty-Five it had caused trouble. Part of the reason for this was that the primary chain was mounted on the left, and the final chain on the right, subjecting the gearbox to twisting loads. Walter concluded that it would be better to continue with the existing Forty-Five unchanged, but design an all-new 45ci or 50ci (750 or 820cc) ohv to run alongside it.

In the end, the idea was dropped because a new steel-strutted piston solved the 74's engine problems, which removed a primary reason for making the 45 ohv at all. As the board minutes recorded: 'At the time this 45 overhead-valve motor was suggested, we were apparently at our wits' end in regard to pistons in our 74 and 80 motors, and it was felt we would have to come to an overhead-valve motor to replace these side-valve motors. Since that time, a new piston has been developed which seems to be the answer to our troubles. If this proves to be correct, the necessity for overhead-valve motors to replace the side-valves is not so great. The one outsanding difference between the overhead-valve motor and the side-valve is that the overhead costs considerably more to make.'

For the 45 ohv, that sounded like the end of the matter, but the discussions went on, and William H. Davidson (son of William A., and the first new-generation Davidson to join the board) sided with Bill Harley. So a proper prototype was built, with new cylinders and heads mounted on a modified 45 bottom end, and in February 1939 it was ridden down to San Antonio in Texas on a 5,000-mile (8050-km) proving run, along with a prototype 74 ohv. There were no major problems, apart from the need for stronger crankcases on the 74 to cope with the extra power. But cost was still the stumbling block, as a 45 ohv would be relatively expensive to build, and the company preferred to concentrate on the 74 ohv and new aluminium heads for the side-valve big twins. Intended to debut in the 1939 line-up, the 45 ohv soon slipped to 1940, then 1941. Finally the war effort put paid to it; but really it was the cost of the bike that made most of the Board so lukewarm about the whole idea.

Fuller, Wyatt (See Razorback.)

Gerberick, Royal
With fellow ex-serviceman Roy Kerle, Gerberick rode a 1920 J-model sidecar outfit 19,000 miles (30,577km) through the Mid-West and Western USA, which was quite an achievement on the roads of 1922.

Gott, Rodney C.
Chief executive of AMF and the man behind the takeover of Harley-Davidson in 1969. AMF's business was mostly based in industrial products, and Gott wished to expand into the leisure industry to balance the company's investment. He was an enthusiast of Harley-Davidson motorcycles and sincerely wanted the company to succeed, but was convinced that rapidly expanding production and non-family management were the answer. As a motorcyclist, however, he did understand the market, something that his successor Tom York (who came from an accounting background) did not.

Goulding
Supplied sidecars, such as the Litecar, to Harley-Davidson for the Forty-Five, which was offered from 1929. Although Harley-Davidson bought sidecars from several sources, and eventually brought their production in-house, it considered at one time buying all of them from Goulding.

ABOVE: Goulding supplied wheels as well as sidecars.

OPPOSITE: Ignoring the fact that this is an Indian bike, it does, however, have a Goulding sidecar, as also supplied to Milwaukee for factory fitment.

Indian

Hadfield, Al
Set a new record for the 'Three Flags' run (Canada to Mexico non-stop), beating Cannonball Baker's run on an Indian. Hadfield, without factory support, covered the 1,689 miles (2718km) on a standard 61ci (1000cc) Harley in 51 hours 22 minutes, a record that was broken by Wells Bennett on an Excelsior the following year. However, Hadfield managed to retrieve it that November, in a time of 49 hours exactly.

Hamm, Fred
Held a number of endurance records for Harley-Davidson before the Second World War. Hamm was a motorcycle police officer from Pasadena who broke records in his spare time. He smashed the Canada–Mexico Three Flags run in 1936, his time of 28 hours 7 minutes on a 74ci (1200cc) Harley-Davidson knocking 10 hours off the existing record. Also notable was his non-stop 24-hour run the following year, using a 61E Knucklehead. A 5-mile circle was marked out with flags and, for night-time running, flares were used on the smooth, dry bed of Lake Muroc. Despite 100-degree daytime temperatures and below-freezing nights, Fred Hamm lapped the circle continuously, covering 1,825 miles (2937km) in the 24 hours. He stopped only for fuel and fruit juice, and the Knucklehead's only problem was needing a new chain, fitted at the 1,400th mile.

Hamsters
Formed over 20 years ago by Arlen Ness, Arlin Fatland and others, the Hamsters are an élite group of Harley custom bike builders. They meet at Sturgis each year and can be spotted by the vivid yellow T-shirts they all wear.

Hardtail
The term for any motorcycle without rear suspension, though it is usually applied to custom Harleys, hence the current Harley-Davidson 'Softail' range, which conceals the rear suspension to give the appearance of a hardtail.

Harley, William A.
Son of the original Bill Harley who succeeded his father as chief engineer on his death in 1943. However, he started with a lowly job in the shipping department, at his father's behest, and learned every aspect of the production process by practical experience. William A. was also an enthusiastic rider, and did some test riding during the development of the Knucklehead. As head of engineering, he supervised Harley-Davidson's post-war model programme, the 7ci (125cc) Hummer, the Panhead and Sportster; but

according to author Harry Sucher, he was a cautious and conservative man when it came to engineering. He retained his job (not to mention the vice-presidency) after AMF took over in 1969 but sadly died in August 1971. His son John was parts and accessories manager.

Harley, William S.
Bill Harley was sometimes described as the 'fourth Davidson', so close was his working relationship with the three brothers. He was a boyhood friend of Arthur Davidson in particular, and the two shared a passion for the new internal combustion engine. William was trained as a draughtsman, so he was instrumental in designing, not only the first Harley-Davidson, but also the bikes in the first 30 years. Naturally enough, he became chief engineer. Bill Harley was a quiet and introverted man, but like the eldest Davidson, he was also a lover of the outdoors and later developed a talent for sketching wildlife as much as for shooting it. He died in September 1943, aged 66.

Harley Owners Group (HOG)
A company-sponsored group of Harley-Davidson riders, it is the first such group in the motorcycle industry and probably the most successful. It was formed in 1983 as a means of communication between Harley-Davidson and its customers via ride-outs, newsletters and details of events. So successful has it been that there are now branches of HOG all over the world. The first year's membership is free with a new bike, though HOG's development as a social group is underlined by the numbers who stay in it after that first year. Seventeen years on, HOG is the world's largest manufacturer-sponsored owners' group, with over half a million members worldwide.

OPPOSITE LEFT
Emblem of the Hamsters motorcycle club.

RIGHT
An aftermarket hardtail frame, just as unforgiving as the original.

LEFT: They get everywhere! A HOG member of the Tenerife Chapter.

Heller Financial
The investment bank that refinanced Harley-Davidson in 1985 after the original financer of the buyout (Citicorp) decided to pull out. According to author Chris Reid, Heller was the only firm that would consider supporting Harley at the time, and without that support, Harley-Davidson would have faced bankruptcy. Heller was impressed by the strength of the brand name, the recovering motorcycle market and the steps already taken to turn the company around. One of Heller's top managers was also a Harley rider, which probably helped.

Hell's Angels
Often associated with Harley-Davidson, especially in their early days, this was a fact that Harley-Davidson not always viewed with relish The first Hell's Angels club was formed in 1948 by a breakaway group from the Pissed Off Bastards in Berdoo, San Bernardino, part of the general growth of 'outlaw' groups in the late 1940s and early '50s, particularly in the wake of events at Hollister, California. Possibly one of the original (certainly the most notorious) motorcycle associations of all time, the Angels now consist of 1,600 members in 100 chapters all over the world. The Hell's Angels thrived as part of the post-war counter-culture, but as a

company, Harley-Davidson did its best to dissociate itself from all the back-patch clubs, seeing it as detrimental to the image of motorcycling which, quite apart from anything else, was bad for business. Instead, Harley publicity material showed clean-cut, clear-eyed, smartly turned out folk riding its bikes: their uniforms would even have passed muster on a parade ground. Not until the late 1970s did Harley riders begin to wear denim and beards in official advertisements. Now, of course, a vaguely Hell's Angel look is *de rigueur* at every HOG meeting and HOG is organized in chapters, just like the Angels.

Sonny Barger was probably the most famous of them all, founding the Oakland, California chapter in 1957. Known as the Chief to his fellow members, he was to spend 21 spells in prison (a total of 13 years) for drunk driving, assaults with deadly weapons, kidnapping, drug dealing and racketeering. He was also, like many Angels, devoted to Harleys, and at 61 years of age still rides one today, a Road King on which he covered nearly 12,000 miles (19312km) in the first three months. He explained his mileage rate was down because of the demands of a book tour to promote his autobiography! (*See also* Hollister.)

High-Flo Muffler
A certain style of tail-pipe fitted in 1934, with the pipe exiting above the lower mudguard stay, and not parallel to the ground, as was traditional.

Hill Climbing
A popular sport in America from the very early years of motorcycling, and particularly vibrant in the 1920s and '30s. Competitors had to ride up an extremely steep slope, and the higher they got, the more likely they were to win, and it was better still if they actually broached the summit. Harley-Davidson was in there almost from the beginning, as big torquey V-twins were ideally suited to this demanding task. Moreover, it had its masters of the art with riders such as Joe Petrali. So popular was it that the factory built many special hill climb bikes in the late 1920s and early '30s. Hill climbing remains a popular sport, though the most successful machines are now lighter-weight off-roaders rather than the big V-twins.

Holiday Rambler
Maker of motorhomes, and bought by Harley-Davidson in December 1986. Harley-Davidson had just gone public, and now had the access to finance (not to

LEFT: Hell's Angels have an extensive network. Here are two English members at the Boothill Saloon.

ABOVE: The High-Flo muffler was a popular fitment in the 1930s.

OPPOSITE: A very American sport: this XR750 has been radically adapted for hill climbing.

FOR 300 MILES
Breaks all World's Records
Ride a Winner
HARLEY-DAVIDSON
HOG

LOBLAN
50

mention managerial freedom) to expand by acquisition. Holiday Rambler promised to even out Harley-Davidson's business by not holding all its eggs in the one motorcycle basket, and by broadening its customer base. That it was able to buy out a company for $155 million, only a year after near-bankruptcy, underlines how rapid Harley-Davidson's turnaround was.

Hollister
The now infamous events at Hollister, a small town in California, on 4 July 1947, have become part of biking folklore. The 'riot' by '4,000 members of a motorcycle club' (*Life* magazine) was undoubtedly blown out of all proportion by the media. More to the point, for the purpose of our story, many (if not most) of the riders arrived on chopped Harleys and Indians. The reporting of this event actually helped kickstart the customizing movement, in which a stripped down, chopped Harley became a symbol of rebellion. It was the 'one percenters' image, which Harley-Davidson did its best to distance itself from over the succeeding 25 years, until it realized that the image (if not the reality) was actually good for business.

OPPOSITE: This clean two-tone Hydra-Glide has a tasteful and restrained amount of extras.

ABOVE: This Hydra, of about the same age, is in its more original form, but has a buddy seat.

Honda
Of all the Japanese manufacturers, it was probably Honda which affected Harley-Davidson most, and for two reasons. One, its advertising campaigns of the 1960s ('You Meet the Nicest People on a Honda') opened up a whole new market for motorcycles, which Harley-Davidson and many of its dealers didn't really understand, and led to the company's retreat from the mainstream into a niche market. Two, Honda's GoldWing struck directly at Harley-Davidson's traditional touring market. It was smooth, quiet and refined, and the fact that Honda later began building it at a plant in Marysville, Ohio shows just how important it was to the US market.

'Hospitals'
An ironic in-house reference to the areas set aside at the York Pennsylvania plant in the 1970s for bikes that needed fixing. At one point, half had missing parts, while others had simply been badly put together. Chief executive Vaughn Beals later recalled that the first 100 XLCR Café Racers off the line in 1976 needed $100,000-worth of work before they could be shipped out.

Hub Gears
Hub gears, mounted in the rear-wheel hub, were used before the gearbox and chain drive as we know them appeared. The earliest Harley-Davidsons had a single-speed direct-belt drive; a hub-mounted clutch was introduced in 1912 (so you didn't have to restart the motor every time the machine was stopped) and a two-speed hub was unveiled in 1914, though Bill Harley had patented a two-speeder four years earlier. But this was only a stop-gap, as a three-speed countershaft gearbox, with chain primary and final drive, came in the following year. The basic engine/transmission layout for Harley-Davidsons would not change for the next 65 years.

Hydra-Glide
Official name given to the first production Harley-Davidson with telescopic front forks, for model year 1949. Although it was no more than manufacturers all around the world were doing at the time, it was a radical step for Harley-Davidson, which had used the same basic design of leading link springer fork since 1908. The new hydraulically-damped forks, with their long travel, soft springing and a soft, wide 5-inch (12.7-cm) section front tyre, added a new dimension of comfort, though diehard traditionalists complained that they lacked the steering precision of the originals. There were also some instances of riders falling off in tight turns as the forks could compress sufficiently to allow the end of the crash bar to hit the ground. A factory kit with firmer damping was subsequently produced which cured the fault.

Hydraulic Valve-Gear
Used by all Harley-Davidsons since the Panhead of 1949. The valves are opened by hydraulic pressure rather than solid pushrods, the advantages being quieter running and lower maintenance.

'I Make the Eagle Fly'
A late-1970s attempt to involve the workforce using exhortations to make them value quality. There was a slight improvement, but at great cost (overtime went up and production dropped), and real quality improvements didn't come until employees were given genuine involvement a few years later.

Import Tariffs
On three occasions after the Second World War, Harley-Davidson asked for higher import tariffs to be slapped on imported motorcycles. The first was in 1950, in response to the increasing numbers of British bikes being sold in America. At the hearing before the US Trade Commission, Harley-Davidson's charge of unfair

HARLEY-DAVIDSON
STOP

competition was scuppered by Alfred Child (*see page 82*), who revealed Harley's then current restrictive practices towards its dealers. As a result, the application was turned down. The same thing happened in 1978 when Harley-Davidson applied for protection against Japanese imports. This time, dealers testified that the Aermacchi-built machines were outdated and uncompetitive against the Japanese. Only in 1983 did the company succeed in getting new tariffs, imposed on imports of over 700cc for five years: Harley had a stronger case this time, and the Ronald Reagan administration was more sympathetic to higher tariffs. In a brilliant piece of PR, Harley-Davidson was able to request that the 1983 tariff be lifted a year earlier than its five-year term.

Indian Dispatch-Tow
Indian's three-wheeled response to the Servicar, which at first sight was seen to have advantages of its own. Even Harley dealers pointed out that the Dispatch-Tow was a handier size than the Servicar, and had an instantly accessible fold-down towbar while the Servicar's had to be stored in the box and bolted on. Harley fitted a fold-down towbar in 1936.

Industrial Engines
Harley-Davidson made stationary industrial engines during the early 1930s. During 1931–32, the plant was operating at a mere 10 per cent of its capacity, and the company desperately needed something other than motorcycles to keep it occupied. Industrial engines promised entry into a whole new market, while making use of existing machine tools. A complete range of singles, horizontally-opposed twins and V-twins was brought out, but sales were poor and the line was dropped after less than two years.

Jack Pine Enduro
A gruelling cross-country competition of around 500 miles (805km), featuring rough tracks to be negotiated and fast-flowing streams to be forded. It seems to have been traditional for younger members of the Davidson and Harley families to take part, and William H. Davidson, later to become president of Harley-Davidson, won the event in 1930.

Japanese Production Methods
(*See* Productivity Triad.)

Jelly Bean System
Nothing to do with confectionary, rather a production method introduced in the 1980s which meant that every different model and colour of Harley would come down the production line, one after another. This direct response to demand from dealers was more efficient than the old method of making the same model in large batches, then storing them until they were sold.

John Brown Motolamp
Name given to the single 7-in (18-cm) headlamp which replaced the twin smaller headlamps in 1913.

Jones Brothers
The first dealer to sell Harley-Davidsons in New Zealand, after a visit by Arthur Davidson in 1920.

Juneau Avenue
Address of the Harley-Davidson headquarters in Milwaukee, Wisconsin.

Kennedy, Pat
Kennedy is best known for his extreme Swedish-style long-forked choppers which he creates at his base in Tombstone, Arizona.

OPPOSITE: The famous Juneau Avenue works, where Harley-Davidson has been based since 1905, has certainly seen some changes.

ABOVE: The Dispatch-Tow was Indian's answer to the Servicar.

RIGHT: Pat Kennedy specializes in long-forked choppers.

Kilbourn Finance Company
A subsidiary of Harley-Davidson, set up in the late 1930s as a finance company for Harley dealers and customers.

Koto Trading Company
Unofficially imported Harley-Davidsons to Japan in the early 1920s, via trader Charles Cable. The bikes were supposedly bound for Mongolia, but Cable thought it more profitable to divert them to Japan, undermining the official importer. (*See also* Child, Alfred.)

Laconia Bike Week
First started in 1923, the event traditionally takes place during the beginning of June. It was based on racing (like Daytona), which had actually been taking place since 1917.

LBO
Leveraged buyout: a technical term for the management buyout of Harley from AMF in 1981. 'Management buyout' is not strictly correct, as most of the finance actually came from banks, not the management itself.

Leading Link Forks
An important early innovation from Harley-Davidson, designed by Bill Harley in 1907. Until then, motorcycle pioneers had mainly used unsprung cycle-type forks, which as speeds rose gave a poor ride and handling; they could also fracture. Harley came up with the solution in his final year at university: two sets of fork legs were connected at the bottom by a pair of forged steel links, which formed a pivot joint. Springing came courtesy of four long coil springs mounted inside the moving set of fork legs – the upper springs for load, the lowers for rebound. The wheel's axle was mounted just in front of the forks' lower attachment (hence leading link). For the low-powered machines of the day, they were rigid, had adequate travel, and a constant wheelbase. And they remained Harley-Davidson's standard front suspension for the next 40 years. They were even reintroduced (albeit in redesigned form) in 1990!

Leather & Lace
A US club and motorcycle association for female Harley riders.

Leno, Jay
Probably the most famous face on American TV, Leno is also well known for his bike collection, which contains a number of Harleys. He regularly acts as 'Road Captain' for the 'Love Ride' each November in California, which raises millions of dollars for charity.

Leonard, Joe
An independent rider in the 1950s, he was not backed by the factory, but nevertheless won the US National Championship on a KRTT. In fact, while the KR was Harley's official racer, the championship was won more often by private riders (notably Carroll Resweber and Brad Andres as well as Leonard) than by factory-backed ones. Between them, Leonard, Resweber and Bart Markel won the title ten times.

Leopold, B.F.
Leopold was instrumental in forcing Harley-Davidson to build its own sidecars. He bought up the Rogers sidecar company in late 1925, Rogers having been Harley's

main supplier, and refused to reserve his entire production for the company, to the point of stopping production altogether. Faced with the sudden withdrawal of its main supplier, Harley-Davidson had no choice but to start making sidecars itself.

Liberty Editions
Special-edition Harleys built in 1985 in honour of the Statue of Liberty and to celebrate its centennary. For each of the 1,750 Liberty Edition bikes sold (there were three versions), Harley-Davidson donated $100 to the Statue of Liberty/Ellis Island Foundation Inc. The company also organized a 'Ride for Liberty', in which Vaughn Beals started from the West Coast and Willie G. took a southern route. They met in Washington and led thousands of riders to Liberty State Park, NJ, where Beals presented the foundation with a cheque for $250,000.

Lightweight Motorcycles
(See Aermacchi.)

Lubrication
In the early days, engine lubrication was taken care of either by a drip feed, by which oil was dripped through a needle valve at a rate that could be adjusted by the rider, or by a hand pump, which one pumped a couple of times every few miles to keep the motor happy. The problem with the latter was that most riders, paranoid that the engine wasn't receiving enough oil, would overdo things and tended to over-oil it instead. The result was oil blow-by past the piston rings, carbon build-up, overheating and premature wear. The only cure was an automatic oil pump, which Harley-Davidson introduced for 1915. Driven off the magneto drive shaft, the pump fed through a sight glass so that nervous riders could reassure themselves that all was well. The pump delivered 4psi, laughably low by modern standards, though at the time Harley-Davidson claimed it would feed against a resistance of up to 70psi. By the standards of the day, it was a precision piece of equipment, and if any pump was faulty, dealers were asked to return it direct to Milwaukee for repair.

Lucifer's Hammer
Famous Harley-powered race bike, which won the Battle of the Twins series three years running, 1984–86.

Magazines
In the early 1970s, the growing custom bike scene spawned a number of lifestyle magazines on the subject, the most famous of which was and still is *Easyrider*.

Magri, Armando
Described by author David Wright as 'the patriarch of Harley-Davidson dealers', he came from Chico, California where he grew up buying, selling and racing bikes when he had the money. He secured a job with Frank Murray's Harley-Davidson dealership in Sacramento in July 1937 and eventually took it over in 1949. His ability to mix with all levels of riders (from the Harley and Davidson families to Hell's Angels) helped business and the dealership grew. Despite the chance to take on Japanese and British franchises, Magri stayed loyal to Harley-Davidson, and retired in the early 1980s, aged 72, leaving a large, healthy business behind.

Marshall & Ilsley Bank
Funded much of Harley-Davidson's early growth via loans, notably a major factory expansion (to 80,000sq ft/7200sq m) in 1911 and again in 1918, when the four Founders reputedly borrowed $3 million from the bank to pay for a huge factory expansion, this time to nearly 600,000sq ft (55,000sq m). Building started in late 1918 and was completed by April 1920. Harley-Davidson had probably the largest motorcycle factory in the world, though to build it the Founders had to overcome their natural distrust of bank loans and the associated interest rates. However, the Marshall & Ilsley Bank was impressed: at the new factory's dedication ceremony, the bank's president declared that Walter Davidson's credit rating was such that he could obtain a million-dollar loan at any time, just by signing his name.

Materials As Needed
System by which only materials immediately needed for production were held in the factory. Harley-Davidson introduced this in the early 1980s in order to cut inventories and costs.

McCann, Jeff
A well respected Californian custom painter, he has worked for many of the top custom bike builders including Arlen Ness, and Ron Simms.

Memorabilia
High prices can be commanded for the most obscure remnants of Harley-Davidson history and there is a large collectibles market for them all over the world.

OPPOSITE
LEFT: Leading link forks on a 1925 J-model – standard Harley equipment until 1948.

CENTRE: A Leather & Lace member.

RIGHT: Jay Leno is one of many celebrity Harley riders.

BELOW: Harley-Davidson memorabilia includes just about everything imaginable.

Military Tricycle (Prototype)
Before the advent of the Jeep, the US military was looking around for something to replace the conventional sidecar outft, and in the summer of 1939 was testing two rival tricycles, both based on motorcycle technology. The Delco company supplied a three-wheeler based on a BMW front end and automotive rear axle, while an Indian trike had already been undergoing testing for about a year. The army was evidently impressed by the shaft-driven Delco, as it afterwards asked Harley-Davidson to submit a bid to build 50 shaft-drive trikes. Bill Harley would have preferred chain drive, and proposed a trike with twin $^5/_8$-in (16mm) chain drives, but this was ignored. But he didn't give up, and later proposed an enclosed rear chain for a 45-based trike, which would overcome all the army's objections to high-maintenance chains.

Nevertheless, the army wanted shaft drive, so that's what it got, and a small batch of 61ci (1000cc) ohv V-twin trikes, with shaft drive and automotive axles, was built in early 1940. But when the first one arrived at Camp Holabird for testing, several problems arose. It vibrated so badly that the fuel line fractured four times; the engine leaked oil, lacked power in top gear and stalled in thick mud; the brakes had to be adjusted often and the whole device was tail-heavy, with a tendency to pop unwanted wheelies, though the latter was acknowledged to be an endemic problem with the tricycle layout. The trike was brought back to the factory and fitted with a slightly bigger 69ci (1130cc) engine, built by modifying a prototype 74 ohv; it was also found that the frame had cracked, so the rear end was strengthened. In the end, the project came to nothing as the army found that the Willys Jeep was far more suited to its needs than any three-wheeler. However, Harley did build an XA-powered trike which was destined to become the new Servicar in 1945.

Miller, Theodore
Assistant sales manager to Arthur Davidson in the early 1920s, and in charge of domestic sales.

Morrow, Carl
Top engine tuner and speed guru. Carl's Speedshop in Daytona is a state-of-the-art high-tech workshop, turning out high-performance parts for road and track Harleys.

Motorcycles As Needed
Dealers protesting against the Bonus Bucks programme in 1983 used this as a parody of Harley's 'Materials As Needed' scheme. They would only order bikes they could sell very quickly, refusing to hold stocks of machines imposed by the factory.

Muhammad Ali (Cassius Marcellus Clay)
Despite once warning Joe Frazier that 'motorcycles is for crazy people', Muhammad Ali later bought a Harley.

ABOVE: Jeff McCann, a leading custom painter.

RIGHT: The 'Mustang' scooter.

Muscular Dystrophy Association
Harley-Davidson's favourite charity, which benefited in particular from the company's 85th birthday to which anyone was invited, as long as they contributed $10 to the MDA. In all, the weekend event raised $500,000 for the charity.

Mustang
Small American moped, the tank of which was used on choppers in the 1970s, coining the phrase 'Mustang-style gas tank'.

Ness, Arlen
The godfather of US customizing, Arlen Ness has been designing and building custom Harleys for around 30 years. He has concentrated on the Low Rider style and also builds outrageous bikes to showcase both his skills and his burgeoning custom parts business, 'Two Bad', for example, being powered by two supercharged 61ci (1000cc) Sportster V-twins. Arlen Ness parts are now an integral part of the custom movement. But though Arlen Ness (and his son Cory, who is part of the business) is thoroughly grounded in Harleys, there are signs of diversification, and the latest addition to the catalogue is a set of Ness-branded race leathers.

New York Stock Exchange
Harley-Davidson went public on the New York Stock Exchange in June 1986, raising $90 million in stock and subordinated debt. Selling company shares on the general market allowed Harley-Davidson to buy Holiday Rambler motorhomes and secured its independence from long-term debt.

Arlen Ness has progressed through just about every custom style, but always comes up with his own ideas. Who else could have thought up this two-wheeled Ford Thunderbird?

Nordberg, George
Production assistant in the early days under William Davidson, Nordberg joined Harley-Davidson in 1909.

NOVA
A new engine programme, launched in 1976, which aimed at producing a complete range of all-new water-cooled engines to power a new generation of modern Harleys. The plan was for a modular system in which a range of engines (including a 30ci/500cc twin, 61ci/1000cc four and 92ci/1500cc six) would all share vital components to cut costs and valves, pistons and gears would be common to all. Parent company AMF seemed willing to underwrite the cost (it was to put $10 million into NOVA), and a contract was awarded to Porsche to develop the range as Harley-Davidson's own engineering department had enough to do, keeping the existing V-twin up with emissions legislation, and developing the upcoming Evolution V-twin. Over the next few years, development continued, and when the Evolution was unveiled in 1983, a few selected dealers were shown a running prototype of the V4 NOVA. If everything had gone to plan, we might have seen water-cooled V4 and V6 Harley-Davidsons on the market by 1985–86.

Unfortunately, AMF's attitude to Harley-Davidson's long-term future changed as the 1970s drew to a close, particularly after new chief executive Tom York came on board. A team of consultants recommended that NOVA get the final go-ahead, at a cost of $60–$80 million. York refused, and instead began to look for ways of getting rid of Harley-Davidson altogether. NOVA was effectively dead from that point on. Ironically, the air-cooled pushrod Evolution proved such a success that the need for NOVA receded

HARLEY-DAVIDSON
XR 750
HARLEY-DAVIDSON
HOG
HOG
1
GOODYEAR

into the background anyway. Only now, nearly 25 years later, is Harley-Davidson thought to be once again working on a new range of water-cooled engines.

O'Brien, Dick
Harley's racing director from 1957, he succeeded Hank Syvertson, being only the third man to hold the post. He had worked for a Harley-Davidson dealer before the war and volunteered for pilot training when war broke out. O'Brien returned to his job as service manager afterwards, before joining Harley-Davidson's racing department which, under his leadership, enjoyed a run of success through the 1960s until 1969, when the AMA rules were changed to allow larger 45ci (750cc) ohv machines to compete with the side-valves. O'Brien's response was the XR750, a short-stroke version of the Sportster V-twin. It was a disaster, but the completely redesigned XR of 1972 had more success, going on to dominate American flat-track racing for the next quarter-century. Dick O'Brien retired in 1983.

OPPOSITE: The XR750, in classic flat-track guise. Once it had been developed and the initial problems overcome, the XR750 came to dominate American flat-track racing. The VR1000 (which you can just see in the left background) faces a tougher task on the circuits.

RIGHT: A 1920s eight-valve racer. After several years of dimissing racing as a waste of good money, the four Founders plunged into competition wholeheartedly with this purpose-built machine.

Ottaway, William
Next to the four Founders, Ottaway was a figure central to Harley-Davidson's racing and engineering policy. He joined in 1913 as assistant engineer to William Harley. He was an instinctive engineer, and in four years designing Aurora Thor machines had made a name for himself in racing circles; Bill Harley realized that Harley-Davidson needed some of Ottaway's natural flair. One of his first jobs was to improve the efficiency of the standard Harley-Davidson 61ci twin, and he also spent much time reducing vibration levels. Ottaway is probably best known for his work on Harley racing machines, notably the II-K (using a tuned version of the 61-ci/1000-cc side-valve twin) and the Model 17 8-valve racer. The latter (after refinement by cylinder-head specialist Harry Ricardo) produced 55hp, plenty to win races in 1915. These technical advances, plus tight team organization from Ottaway, allowed Harley-Davidson's all-conquering Wrecking Crew to sweep the racing boards in the late teens. It was also William Ottaway who warned Bill Harley that his clutch design might infringe Bendix patents and was proved right when Bendix took Harley-Davidson to court. He increasingly concentrated on road bikes after Harley-Davidson dropped its big racing programme in 1921 and Joe Petrali took on the development of race machines. When William Davidson died in 1937, Ottaway took over as interim production manager and agreed to allow trade unions into the plant for the first time. In 1943 he took over as chief engineer when Bill Harley died, though this function was soon passed on to Bill's son, William.

Pacific Motors
This is a story of shady goings-on in 1920s Los Angeles, where Indian dealer C. Will Risdon had the upper hand, particularly in the lucrative and high profile police sales. He decided to retire in 1925, and looked around for someone to buy his well-established, well-equipped dealership. An offer came in from A.F. Van Order but Risdon was suspicious, knowing that Van Order worked for Harley dealer Rich Budelier; but Van Order's story was that he had been sacked, and wanted to set up in competition with his old boss. Risdon was persuaded, and his dealership passed to Van Order's holding company, named Pacific Motors. He immediately increased the fire insurance coverage to its maximum, and the following month the entire premises burnt down! Nothing was ever proved in court, but on the night of the fire Van Order was 'out of town' and Budelier was found in bed, fully clothed and with his shoes on.

Parker, Scott
An official Harley-Davidson racer in the 1980s/90s and the youngest rider ever to earn an AMA expert licence. He was team-mate to Jay Springsteen in the official racing team for 1986 and won the National Championship himself in 1990 and '91.

Parkhurst, Leslie
'Red' Parkhurst was a central member of the Harley-Davidson Wrecking Crew, and a highly successful and popular racer on both side-valve and eight-valve Harleys. He started racing Harley-Davidsons privately, serving his apprenticeship on the notorious pine-planked Motordromes (nicknamed 'Murderdromes') but caught the attention of the factory team when racing at Milwaukee. He was recruited to the team and, being recently married, abandoned the Motordromes for safer dirt tracks. The following year, after many wins, he had a serious accident at Rockford, Illinois, but carried on racing until the USA entered the First World War. Despite declaring he would never return to racing, Bill Ottaway managed to persuade him back with the lure of a ride on the eight-valve racer for the 1919 season; he won the 200 Mile Internal Road Race Championship at Marion, Indiana and in early 1920 set up new speed records (although not officially accepted) of 112.61mph (181.2km) over the mile on an eight-valve and 82.09mph (132.1km) on a sidecar outfit. But by this time Red Parkhurst was one of the oldest racers on the track and he decided to retire for good, citing his responsibility to his wife and child.

Perkins, Dudley
An early Harley-Davidson dealer in San Francisco, who helped Tom Sifton obtain his sub-dealership in 1929.

Pershing, General John
Led the first US military operation to use motorized units, including Harley-Davidson solos and sidecars in the 1916 attempt to track down the army of General Francisco Villa in Mexico.

Petrali, Joe
Almost single-handledly took on Harley-Davidson's minimal competition strategy

Scott Parker, US National Champion in 1990 and '91 on an XR750.

from the late 1920s. He was a terrific rider on both tracks and hill climbs (Hill Climb Champion 1932–38), a practical mechanic, and was central to the development of the 61E Knucklehead. Born in 1904 in San Francisco, the son of Italian immigrants, Petrali made his name racing in the 21ci (350cc) class, and was taken on by Harley-Davidson to 'racerize' its own 21ci Peashooter. While helping develop the Knucklehead, he warned the company that the bike's oil leaks and valve spring breakage problems needed sorting out before production. He also set up speed records with the Knucklehead. With Hank Syvertson, he built a streamlined bike with twin carburettors, higher compression, different cams and magneto, plus a high top gear that allowed a theoretical top speed of 160mph (257km). On the first run, the bike went into a serious tank slapper as Petrali approached 100mph. He came in and ordered that all the aerodynamic bodywork be removed, then went out again and captured the mile speed record at 136.15mph (219.1km). This was disappointing for Petrali, as Walter Davidson had promised him a $1,000 bonus if he got to 150mph! Joe was evidently given many of the odd jobs at Harley-Davidson, and in 1938 supervised the dismantling of a Crocker V-twin, which Milwaukee suspected of encroaching on Harley-Davidson patents. He resigned from the company that year after Walter Davidson refused him a job in the experimental department and went on to work for Howard Hughes, later taking to the speaker circuit where he was able to air his reminiscences of Hughes and Harley-Davidson. Joe Petrali died in 1974 as a result of heatstroke suffered at Bonneville Speed Week.

Pinehurst Meeting

This was a seminal week in Harley history. Vaughn Beals, the man brought in by AMF to turn Harley-Davidson around, decided in April 1976 to take all Harley-Davidson senior managers on a week-long brainstorming session to Pinehurst, North Carolina where the long-term project strategy could be thoroughly explored. What emerged was a two-point plan. First was to improve the existing V-twin gradually and this was eventually unveiled as the Evolution engine, which carried Harley-Davidson right through the 1980s and '90s. Second was the NOVA family of all-new engines. (*See* NOVA.)

Poetry

Martin Jack Rosenblum may well be the only Harley-Davidson poet in the world. He has a doctorate from the School of Objectivist Poets, and in his 40s achieved a lifelong ambition to own a Harley. While recovering from a near-fatal illness, he made a complete career change from teaching poetry at university level to helping students lacking in college skills. He now works as Harley-Davidson's official archivist at Juneau Avenue. Here is a selection of his work, all taken from his book *The Holy Ranger*, and all are produced with the kind permission of the man himself.

The Name Is Harley

standing there with the look
of invention in front of that
original shack new ideas took
hold which built a reputation
all the way into the American ideas

that govern individuality & style:
William S. Harley has lent his name

to the modern horsemen whose
apocalyptic ride strengthens
open trail visions that make

this photographic action take place
this sunrise in April as I take the
keys to my Harley machine then walk

the irregular stone path to
the garage contemplating in
poetic anticipation exactly
what will
be coming
within an
afternoon –
upon this
Iron Name

Dr Martin Rosenblum is probably the world's only Harley-Davidson poet. He's certainly the only one sanctioned by the factory!

Evolution Amulet

the 30th Anniversary XLH sat in the dining room
during winter months & once after a late movie
when all slept & the first real snow landed
I got on it & turned the lights toward a
front window shining into the blanket
being tossed upon the birch tree
as winter's comfort & signalling
with yellow flashes reflecting
on an empty television screen:
the first spring rain bringing
this Sportster down the ramp
& across the lawn & sidewalk
to the garage where it
fires up once the carb
swallows ample passion
but the rain keeps any
travel in the alley to
just beads upon fresh
wax or upon the chrome my reflection
while I get down nearer the exhausts
to hear what winter kept in silence

Harley-Davidson 85th Anniversary Homecoming Contemplation

out of the Kenosha holding strip
onto 94 with all areas alongside

covered by raised fists & victory
then on p43 of the HD magazine I

find my Harley racing orange front
fender in that photo coming around

the bend in back & my beard sprouting
out from the black helmet – well being

there in June & here now in my office
between appointments just from buying

this to read & opening my napkin
to keep sandwich crumbs off that

front fender it coming at me
as I was upon it with cycles

behind it columns roaring
& such lovely ladies with

leather fringe sparkling
upon the interstate path

into Milwaukee Iron
coming home I rode:

a pack of Strokers
at my side & front

to rear passion
burning V-twins

in the mystical
gas tank logos'

signatures that
chant the poems

I cannot write

Mid-Summer on a Harley-Davidson

Wisconsin cornfields strangled
without water

& gravel shoulders coughing
dust in the wake

of that eighteen-wheeler ahead
so the downshift & then faster

onto this county trunk
as blackbirds' chatter

stick in the windless
hum of tires going to

the broken yellow
lines as a center

for the holy
morning ride

History's American Springer

all the imports hid behind plastic
fairings and their riders crouched
on top of them wearing bright
tennis shoes
in the later
eighties when exposed satisfaction
aesthetically sprung onto those
Softail front ends in chromed
memories of the
future here in an
American reply with
our riders deep inside
of these machines such
that every element
upon Milwaukee
street poetry was concretely observed
through that
Springer expanding & contracting
anytime the rider's heart would &
blood pulsed through the sparkling
coils straight from homemade poetic

Potato Patent
Harley-Davidson has been famously litigious over the use of its trademark, most notably of all over the attempt to patent the 'potato-potato' exhaust note. This is a distinctive sound, a product of the 45-degree V-twin which because it's not produced by any other motorcycle is instantly recognizable as emanating from a Harley. The company reasoned that this too could be copied by a rival manufacturer, and so began exploring whether it was legally possible to patent a noise, even one as distinctive as this. The legal process went on for several years, and in 2000 an Australian musician by the name of Bill Cook was actually to claim that he'd taken out an American copyright on potato-potato when making an album called *Steel Stallions* in 1993. According to Cook, legal counsel for Harley-Davidson had been trying to reclaim ownership of the patent through the US Patent and Trademark Office, despite opposition from rival manufacturers. The musician cheekily maintained that he was prepared to hand the copyright over if Harley-Davidson gave away a copy of his album with every bike it sold, for which he would charge $15 a time. It's true that Harley-Davidson gave up its attempt to patent the sound in 2000, but was it really because of Cook? 'It's a nice story,' said one Harley-Davidson official, 'pity it's not true.'

Presley, Elvis
The King was a confirmed Harley rider, a fact which made it to the cover of *The Enthusiast* (Harley's own magazine) in May 1956. According to the text, he had started out on a 165 and was by then onto his third Harley, a brand new red-and-white KH. It was, of course, great publicity for the company, so it must have been disappointed by the1964 film *Roustabout*, in which Elvis rode a Honda!

Price Fixing
Price fixing was a feature of the relationship between Harley-Davidson and Indian between the World Wars. It seems astonishing now that these two firms, which were such bitter rivals and which indulged in cut-throat competitive techniques, should also meet once a year to agree on common prices. The first meeting was in 1922 between Arthur Davidson and Indian's general manager, Frank Weschler. In the words of author Harry Sucher, these long-term rivals shared 'crêpes Suzette and lobster thermidor in an atmosphere of somewhat strained conviviality'. It was agreed that henceforth, similar competing models from Harley-Davidson and Indian

Despite riding a Honda in the film *Roustabout*, Elvis owned a Harley. It was an early XL Sportster, with screen, and appeared (as seen here) on the cover of *The Enthusiast* magazine.

THE Enthusiast
A MAGAZINE FOR MOTORCYCLISTS

would carry the same prices. Typical was the 1931 dinner meeting at New York's Waldorf Astoria Hotel: the Harley-Davidson Big Twin and Indian Chief would both cost $320, the Harley-Davidson 45 and Indian Scout, $295, and so on. At one point, Indian president, Paul du Pont, turned up unannounced, and warned of the consequences if the contents of these secret meetings ever leaked out. Price fixing was of course illegal.

Productivity Triad

Three things crucial to Harley-Davidson's survival and turnaround in the 1980s were employee involvement, just-in-time production and allowing workers to control quality for themselves. In 1981, Harley-Davidson's central problem wasn't a lack of demand for its bikes, but the way they were made; in other words, inefficiently, expensively and of poor quality. The answer was to use Japanese production methods, of which the Productivity Triad was Harley-Davidson's interpretation.

1) Employee Involvement: There was a wealth of knowledge and experience on the shopfloor in Milwaukee and York, but it was being ignored. This came to a head in the AMF days, when production was forced up despite what those on the shopfloor said. Quality circles had already been tried in a half-hearted fashion, but this time they worked, because they (and other employee involvement techniques) had the commitment of management, and employees realized that they really were being heard and could have a genuine influence.

2) Just-In-Time (JIT) Production: Traditionally, like much US industry at the time, Harley-Davidson had relied on large batch production of both certain components and particular models of bike. This meant holding large stocks of raw materials and parts, sometimes for months until they were needed, which tied up a great deal of capital and slowed down the production process. Instead, JIT delivered only those parts to the assembly line which were needed to build each bike (*see* Jelly Bean System). Meanwhile, departments were reorganized to enable them to rapidly switch between making small batches of different parts, rather than large batches. In the ten years after JIT was introduced, Harley-Davidson's inventory was reduced by 75 per cent, scrap and rework was down by 68 per cent, and productivity up by 50 per cent.

3) Statistical Operator Control: A much simpler idea than it sounds. Employees were given the training to monitor the quality of their own work. Once they were proficient in this, problems could be rectified at source as soon as they arose because they were detected immediately by the person doing the job, rather than by a quality inspector further down the line after 500 faulty parts had been made or (worse) by no one at all.

These three methods were all inter-related, and it is doubtful that any one of them could have worked without the other two. Together, they transformed the efficiency and effectiveness of Harley-Davidson's production process, and turned the company around.

Quality

(*See* Productivity Triad.)

Rat's Hole

Big Daddy Rat holds these world-famous custom bike shows at Daytona Bike Week in March and the Sturgis Rally in August. He also puts on an annual event at the Essen Motor Show in Germany.

Razorback

A make of custom Harley during 1993–94.

OPPOSITE: A selection of customs at the Rat's Hole Show, at Daytona in 1989. These sleek 'bodywork bikes' are a phenomenon of the 1990s.

RIGHT: Rat's Hole, 1990, and Big Daddy was there in person.

Built by Wyatt Fuller (who has been riding for 30 years but only customizing for three), the Razorback Motor Works specialized in clean-looking single-colour bikes, with the engine exposed but all the nooks and crannies smoothed out and filled in. So successful was the look that Wyatt was bought out by Harley-Davidson in 1994 and the man himself was contracted to design accessories.

Reagan, Ronald
As well as agreeing to a new import tariff in 1983 to protect Harley-Davidson, President Ronald Regan toured the York assembly plant in May 1987. All politicians like to be associated with success, and Reagan was no exception, particularly as the tour followed Harley-Davidson's high profile request that the tariff be dropped a year ahead of schedule. It was a symbolic moment: at one point, Reagan was invited to start a newly-built Sportster: 'This thing won't take off on me, will it?' He never was a motorcyclist.

Red River Run
On the annual memorial weekend, the Vietnam Veterans Motorcycle Club gather in Red River, Mexico to ride, party and remember their fallen comrades at the Vietnam memorial chapel in the mountains.

Remy
Remy electrical equipment was introduced on the 61ci (1000cc) J-model in 1916, in the form of battery and lights. In 1919, the factory dropped these in favour of its own electrics, though it was over a year before they were acceptably reliable.

Resweber, Carroll
An official factory rider in the late 1950s who won the National Championship four times, in 1958, '59, '60 and '61. Incredibly, he did all this while working a full-time day job for Mercury Marine. Not surprisingly, his home town of Cedarburg, Wisconsin, held a 'Carroll Resweber day'. He was on course for a fifth title when he ran into Jack Gholson and Dick Klamforth at the Lincoln, Illinois track in 1962 where the dust was so thick that he reputedly had his eyes shut at the time. He was seriously injured, lying unconscious for nine days, and was left with one arm and leg shorter than the others. But Resweber survived and went on to work in the Harley-Davidson race department. One of the better Resweber anecdotes came out of a race in late 1950s Laconia. It was early days for the racing K-model, and the gearbox was turning out to be too weak. Resweber didn't want to race that day, having a huge hangover, but being a factory rider he had to do so. So he determined to break the K's weak gearbox and retire early. However, and despite full-throttle clutchless heavy-footed gearchanges, it ran the full distance.

Ricardo, Harry
An English cylinder-head specialist, whose pioneering work on four-valve combustion chambers inspired generations of engineers. At Joe Petrali's suggestion, he was brought in to develop Harley-Davidson's eight-valve racer for 1915. Walter Davidson refused at first, faced with the cost of Ricardo's steamship ticket and his fee, but relented when Petrali

SALOON
BULL O' THE WOODS
Satellite
Big Screen T.V.

reminded him that $25,000 had already been spent on the the eight-valve. Ricardo duly worked his magic and the bike was transformed into a race winner. Less successful was the side-valve Model A 21ci (350cc) single that Harry Ricardo had had a hand in and which had a tendency to seize if ridden hard for any length of time.

Rikuo
Model name for Harley-Davidsons built under licence in Japan in the 1930s. (*See* Child, Alfred.)

Robert Baird & Co
The Wall Street firm of brokers which launched Harley-Davidson shares on the stock market in 1965. Until then, much of the company was still directly owned by the Harley and Davidson families, 50 members of the two clans owning 53 per cent of the voting shares and in effect controlling the company's destiny. But by the mid-1960s, Harley-Davidson was losing market share and desperately needed money. One way to raise it was to sell shares on the open market. Robert Baird & Co was later retained to advise Harley-Davidson on the AMF takeover.

OPPOSITE: Scene at the Red River Run.

RIGHT: A Shriner Servicar.

Robinson, Dot
National president of Motor Maids of America in the 1940s and wife of Harley-Davidson dealer Earl Robinson.

Robinson, Earl
Detroit Harley-Davidson dealer from the 1930s, who set transcontinental records solo and with his wife Dot. He rode the 3,000 miles (4828km) from New York to Los Angeles on a tuned 45 in 3 days 6½ hours in 1935, then made the same run with Dot, setting a sidecar record.

Rogers Company
Supplied Harley-Davidson with sidecars up to 1925, when founder B.F. Rogers died and the company was bought by Eros Manufacturing. Rogers always resisted Harley-Davidson's demands for it to make sidecars for it exclusively, and carried on selling to other manufacturers such as Ace and Excelsior. However, Harley-Davidson did persuade Rogers to build a special wide-bodied sidecar which would carry two adults. Its 56½in (143.5cm) track was the same as the standard US car track, though it was designed more with exports in mind rather than the home market.

Rourke, Mickey
Yet another high-profile celebrity Harley rider of the 1980s. Screen actor Rourke spoke of his fascination with Harley-Davidson: 'It's a personal thing that can't be described. It's part of you.' Not that Hollywood Harleys were a new idea: in the 1930s and '40s, stars like Clark Gable, Robert Young and Tyrone Power were all happy to be photographed astride big V-twins. On the other hand, the Harley as a celebrity fashion accessory may have gone a little too far, for it's been said that in Hollywood, Harley-Davidsons are just like a certain part of the human anatomy – everyone had one.

Scherer, Julian
'Hap' Scherer was Harley's publicity manager in the early days, having joined the company in 1915. He was a particularly strong exponent of the Sport Twin, realizing that the whole American motorcycle industry was in danger of only catering to the limited enthusiast market, with big V-twins. Scherer, along with other industry salesmen, wanted to expand the market with easier to ride machines like the Sport Twin. In 1920, he actually rode one from Denver to Chicago in 48 hours, but sales fell off after he left the company. He was sacked by Walter Davidson after allegedly co-operating with Harley-Davidson's main rivals. Scherer later went to work for Rich Budelier's new Harley-Davidson dealership in Los Angeles.

Schwinn, Ignaz
The bicycle magnate who bought up Excelsior, for some time the number three American manufacturer, after Harley-Davidson and Indian, and kept it going until 1931.

Servicar
The three-wheeled Servicar used the front end of a 45 married to a two-wheel rear end and a load-carrying box. Originally intended for mechanics to collect vehicles for repair, it proved ideal for traffic police and meter maids, becoming something of an American institution.

Shriners
Believe it or not, there is a strong association between that peculiarly American phenomenon, the Shriners, and Harley-Davidson. Drill teams have become common, and perform intricate, highly

disciplined manoeuvres at outdoor shows. Many of them are based in the US's south-eastern corner, and the Al Menah team from Nashville is typical. It has 16 members mounted on identical police-style Harleys (without the sirens or red lights) and was formed in 1958. Even by the early 1980s, it had still won more top three trophies than any other team. The competitions aren't just about bike control; there's a strong militaristic aspect to the Shriners in their demand for cleanliness, uniformity and a good turn-out. Inspections are said to be just as tough as those for the Marines!

Side Van

Introduced in 1915 as an adjunct to the successful sidecar line. It was basically a wooden box on a sidecar chassis, with a hinged top for loading, and proved very popular with commercial users. Among the factory options was your own

ABOVE: Before the Servicar, side vans (commercial sidecars) were popular with businesses that couldn't afford a Ford van.

LEFT: Silent Gray Fellow undergoes silent contemplation: very few of these bikes are now unrestored.

OPPOSITE: Big man, big bike: Ron Simms of Bay Area Custom Cycles.

company's signwritten name on the box, at 10 cents per letter, and a folding top and side curtains to protect the rider from inclement weather.

Sifton, Tom
An almost legendary tuner of Harleys, Tom Sifton started out as a sub-dealer in San Francisco in 1929 and became a full dealer in 1933: but he increasingly spent his time tuning Harley-Davidsons for competition. With detailed work on the porting and valves, he was able to make the unpromising 45 into a race winner, but just how he did it remained a secret from fellow tuners and even from the factory's racing department. In fact, Sifton went on making side-valve Harleys go faster for decades, working his magic on the WR racer in the late 1940s and the KR in the 1950s and 1960s.

Silent Gray Fellow
The nickname given to the early Harley-Davidson single, owing to its standard colour and quiet running. It was a typical machine of the time, with belt drive, inlet-over-exhaust head and atmospheric inlet valve. Top speed was around 45–50mph (72–80km/h) and the engine's peak was about 500rpm: only when more positive valve operation was used could engine speeds rise. But of the thousands of similar machines which appeared in both America and Europe in the first decade of the 20th century, the Silent Gray Fellow was one of the few survivors, for no better reason than that it was well made and reliable.

Simms, Ron
Has operated as Bay Area Custom Cycles in San Francisco for 30 years. His tough 'gangster'-style customs with distinctive skeleton/skull paint jobs are instantly recognizable.

Spiegelhoff, Johnny
A racer in the 1930s and '40s, and though better known as a Indian devotee, Spiegelhoff was a committed Harley rider in the 1930s who worked as an inspector on the Milwaukee assembly line. But he switched allegiance when the Harley-Davidson racing department refused to make the changes he wanted to his bike. So he left his job, bought an Indian, and scored several race wins, notably at Daytona in 1947. He even took on an Indian dealership in Milwaukee, right on Harley's home turf.

Springsteen, Jay
A legendary racer on dirt or pavement, his career stretches back over 20 years and continues to this day.

Sturgis Bike Week
(See Black Hills Classic.)

Styling
The styling of Harley-Davidsons – the way they look – is now central to their appeal, and just as important as the trademark V-twin. But a styling department wasn't established until 1963, when president William H. Davidson phoned his son Willie G. (then working for an industrial design company) and invited him back to the family firm to run it. Even so, it was eight years before Willie G's first seminal bike appeared, the Super Glide, using an Electra Glide engine and frame with a lighter-weight Sportster front end. It was a nod in the direction of contemporary customizers, and the first true factory custom. The Super Glide was followed by the less successful Café Racer (whose European look was too remote from the traditional Harley) and the very successful FXS Low Rider, with raked, lengthened forks, low seat and fat rear tyre. The Low Rider sold by the thousands, but the clever part was that, like most of Willie G's designs, it used mainly existing components to save money and time. Since then, other styling milestones have been the Wide Glide (with wide-spaced forks – another custom favourite) and the Softail (with hidden rear suspension to ape the authentic hardtail look). The Springer Softail used an updated version of Bill Harley's original Springer front forks and the Fat Boy (another good seller) has solid wheels and a single colour. As the 1990s ended, Harley-Davidson seemed to have reached the point where the way its bikes looked was far more important to buyers than any technical considerations.

SuperRide
A nationwide demonstration ride programme in 1985 which took place over three weekends when 40,000 potential customers took 90,000 test rides, though not enough bought bikes to justify the $3 million that SuperRide cost to set up. But since then, Harley-Davidson has become one of the foremost companies for offering demo rides at shows, events and dealerships.

Syvertson, Hank
Long-time member of the Harley-Davidson racing department, who assisted Joe Petrali in the development of the Knucklehead and in building Joe's record-breaking version of the same bike.

LEFT: They don't come much more experienced than this. Here Jay Springsteen is lost in contemplation before a race.

OPPOSITE: The XLCR Café Racer was one of Willie G's rare flops. Call it a European-type Sportster or what you will, but few were sold.

Teerlink, Richard
Harley-Davidson's chief financial officer at the time of the 1984 crisis with Citicorp. Citicorp, Harley's original financial backer from the 1981 buyout, wanted its money back; so Teerlink, together with Vaughn Beals, Tom Gelb and others, managed to negotiate a refinancing deal with Heller Financial. It was finally signed on 31 December 1985, just weeks before the company would have needed to file for protection against bankruptcy, for which a legal team had already been hired. Richard Teerlink later succeeded Vaughn Beals as chief executive officer.

Terminator
Who else but Arnold Schwarzenegger could plays an android, returned to our time from the future in order to protect the future saviour of the human race? And, of course, he rides an all-black Harley-Davidson. As he says to the prone biker he's pointing a large gun at: 'Give me your clothes, and your motorcycle.'

Thompson, Charley
One-time president of Harley-Davidson, who once said that the biggest job for dealers was to find things for customers to do with their bikes once they'd bought them. He was reflecting the way the market was changing into a leisure industry, with accessories, clothing and dealer-organized events, like ride-outs, being at least as important as the motorcycles themselves.

Tipton, Ken
Builder of the Harley-Davidson specials from the 1950s, who went on to produce his own proprietory tuning parts.

Tritten, Ray
Brought in by AMF in 1973 to tackle Harley-Davidson's continuing problems. The moving of bike assembly to a factory in York, Pennsylvania had created its own problems, while quality and market share were heading downhill. He made some small improvements to production and forecasting, but probably his biggest contribution was to bring in Jeff Bleustein (*see* above) to head up a new, more professional engineering department and to appoint Vaughn Beals as AMF's man in Milwaukee. Beals was a particularly good choice, as this finally put a senior AMF person in day-to-day contact with Harley-Davidson management. Previously, AMF had been attempting to control both Milwaukee and York plants from its own remote headquarters.

Triumph
Triumph, the famous manufacturer based in the English Midlands, was a thorn in Harley-Davidson's side in the 1940s, '50s and '60s. It was probably the most successful of the makes imported into the US after the Second World War, notably with the twin-cylinder Thunderbird and Bonneville. Although smaller and lighter than any Harley, the Triumphs were faster and more nimble, and by 1950 imports had captured 40 per cent of the American motorcycle market. Harley-Davidson's response to Triumph and the other imports

was slow to come, but arrived in the form of the side-valve K, overhead-valve Sportster, and later with the smaller Aermacchi bikes. Harley-Davidson's official view was that none of these imports were direct competitors to its own big V-twins, a short-sighted view that nearly led to its demise. Ironically, Triumph and its British contemporaries used much the same argument when confronted with Japanese competition.

T-Shirts
T-shirt slogans featuring Harleys have become something of an art form. Here is a selection:
'God Rides a Harley'
'Harley-Davidson: If I had to explain, you wouldn't understand'
'Harley's Best, **** the Rest'
'I'd rather see my sister in a whorehouse than my brother on a Honda'
'On the Seventh Day, God Created the Harley-Davidson (And on the eighth, he traded it in for something lighter and faster)'

Tucker, Mike
One of the first people to drag race a Harley. Tucker used a more or less standard FLH with a tuned engine and all the superfluous equipment removed.

Vietnam Vets (*See also* Red River Run)
A US national club whose members all served in the Vietnam conflict from 1964–73. They all sport patches commemorating their fallen brothers and others declared missing in action.

Von Gumpert, Eric
A long-standing export manager from 1913, Eric von Gumpert negotiated the first exports of Harley-Davidsons to Britain, and was involved in many of the early export markets, including Europe and Japan. As a veteran employee, he prepared Harley-Davidson's submission to the Tariff Commission in 1951, applying for higher import duties on bikes brought into the USA.

Walker, Otto
Captained the Harley-Davidson racing team in 1915, winning the 300-mile (483-km) race through Los Angeles' closed streets at 68.31mph (110km/h). Team mate Red Parkhurst came second.

Warbird
A special body kit designed by Mike Corbin of Hollister, California, which can be fitted to big twins or sportsters giving them the appearance of fighter aircraft and a racier image. They come unpainted, the builders, in this case Arlin Fatland, getting individual paint jobs designed.

Warr, Fred
A famous Harley-Davidson dealer in London, who later became the official importer for Britain. He is still in business as F.H. Warr & Co.

Watson, Duncan
Official Harley-Davidson importer for Britain from April 1914, though business only really got underway after the First World War, in 1919. He was also Lord Mayor of London and was knighted by George V. Watson made a good job of establishing Harley-Davidson in Britain, but withdrew in 1925 when a new 33 per cent import tax made the American bikes prohibitively expensive. It would be many years before Harley-Davidson had another UK importer.

Web Sites
There are probably thousands of web sites associated with Harley-Davidson, but here are four that are official:
www.harley-davidson.co.uk
www.harley-davidson.com
www.hog.com
www.buell.com

The Wild One
Stanley Kubrick's 1954 film, starring Marlon Brando, was supposedly a dramatization of the events of 1947 in Hollister, California (*see* above), which reinforced the bad boy biker image in the minds of the general public. The film was actually banned for a time, until 1970 in England, but the image stuck. It really didn't matter to most people that Brando rode a Triumph – in the eyes of the bikers, Lee Marvin's rival gang all rode chopped Harleys, so it followed that they must be the *real* outlaw bikes.

The Wrecking Crew
The almost legendary Harley-Davidson racing team of lthe late 1910s and early '20s, which at that time swept the board in America. Under the direction of Bill Ottaway, a hugely talented group of riders was brought together, with strict team

tactics and pit-stop techniques adding to the success. Prominent riders were Jim Davids, Ralph Hepburn, Walter Higley, Fred Ludlow, Otto Walker and Ray Wishaar.

Zimmerman, Fred
A dealer in Rochester in the 1920s, Fred Zimmerman enjoyed the disinction of selling new Indians under the 'Rochester Harley-Davidson' name. Fred obtained a Harley-Davidson franchise for the area in 1922, despite being a Jew. (The Founders were basically anti-Semitic, and had a policy of excluding Jews from any part of the business.) Zimmerman's new dealership did well, particularly with the advent of the fast Two Cam Harleys. It was said that he used to fettle Two Cams used by the booze runners in those Prohibition days, as well as those used by the state police to chase them! The story goes that a business rival informed Harley-Davidson of Fred Zimmerman's ethnic background and factory support for his dealership was gradually withdrawn. In the end, Zimmerman cancelled his franchise and took one with Indian instead, but kept the name 'Rochester Harley-Davidson'. His right to do so was upheld in court.

OPPOSITE:
LEFT: Triumph twins (this is a flat-track racer) seriously threatened Harley-Davidson in the 1950s and '60s.

RIGHT: However, some very interesting people ride Harleys nowadays.

RIGHT:
Corbin Warbird, as painted by Arlin Fatland.

YEAR ON YEAR

All years refer to Model Years.
Listed are main models for sale only.
Prices are retail for US market.
Main specifications and changes only are listed,
not minor year-to-year changes

1903

Model: 24.7ci (405cc) F-head single
Price: $200
Production: 3
Colour: Black with Gold pinstriping (by Aunt Janet Davidson)
Specification: 24.7ci F-head single, 3 x $3^{1}/_{2}$in (76 x 89mm) bore x stroke, auto inlet valve, belt transmission, loop frame, 28in-wheels (711mm), wheelbase 51in (130cm), weight 185lb (84kg)

1904

Model: Model 0, 24.74ci (405cc) F-head single
Price: $200
Production: 8
Colour: Black with Gold pinstriping
Specification: As 1903

1905

Model: Model 1, 24.74ci (405cc) F-head single
Price: $200
Production: 16
Colour: Black with Gold pinstriping
Specification: As 1903

1906

Model: Model 2, 26.8ci (440cc) F-head single

A 1905 single. Few changes occurred that year, but production doubled.

Price: $210
Production: 50
Colours: Black with Gold pinstriping, Renault Grey with Red pinstriping
Specification: 26.8ci (440cc) F-head single, $3^{1}/_{8}$in x $3^{1}/_{2}$in (79 x 89mm) bore x stroke, optional hand crank for engine, otherwise as 1903

1907

Model: Model 3, 26.8ci (440cc) F-head single
Price $210
Production: 150
Colours: Renault Grey with Red pinstriping, Black
Specification: Sager front fork used, otherwise as 1906

1908

Model: Model 4, 26.8ci (440cc) F-head single
Price: $210
Production: 450
Colours: As 1907
Specification: As 1907

1909

The Range
Model: Model 5 - 30ci (**500cc**) F-head single, 28in (711mm) wheels, battery ignition
5A - 30ci F-head single, 28in wheels, magneto
5B - 30ci F-head single, 26in (660mm) wheels, battery ignition
5C - 30ci F-head single, 26in wheels, magneto
5D - 50ci (819cc) F-head V-twin, magnet
Prices
5 - $210
5A - $250
5B - $210
5C - $250
5D - $325

LEFT: There was little difference between the 1905 (front) and 1907 (rear) singles.

TOP: The 1907 Model 3 with Sager front fork.

ABOVE: A rare 1909 5D V-twin, only produced in that year.

Production
5 - 864
5A - 54
5B - 168
5C - 36
5D - 27
TOTAL - 1,149
Colours: As 1907
Specification:
5-5C - 30ci F-head single, $3\frac{5}{16}$in x $3\frac{1}{2}$in (78 x 89mm) bore x stroke, auto inlet valve, Bosch magneto on A and C.
5D - 50ci (820cc) F-head V-twin, 3 x $3\frac{1}{2}$in (76 x 89mm) bore x stroke, auto inlet valve, drip-feed lubrication, 7hp, Bosch magneto, belt drive, loop frame, 28in (711mm) wheels.

1910

The Range
6 - 30ci (**500cc**) F-head single, 28in (711mm) wheels, battery ignition
6A - 30ci F-head single, 28in wheels, magneto
6B - 30ci F-head single, 26in (660mm) wheels, battery ignition
6C - 30ci F-head single, 26in wheels, magneto
6E - Racer, 30ci F-head single
Prices
6- $210
6A - $250
6B - $210
6C - $250
6E - $275
Production
6 - 2,302
6A - 334
6B - 443
6C - 88
6E - 1
TOTAL - 3,168
Colours: As 1907
Specification
Belt idler fitted, and wider $1\frac{3}{4}$in (44mm) drive belt, otherwise as 1909.

TOP: The 1910/11 single: an option of 26- or 28-inch wheels allowed different seat heights.

ABOVE: The 1912 X8A combined a 30ci (500cc) single with magneto and rear-wheel clutch.

1911

The Range
7 - 30ci (**500cc**) F-head single, 28in (711mm) wheels, battery ignition
7A - 30ci F-head single, 28in wheels, magneto
7B - 30ci F-head single, 26in (660mm) wheels, battery ignition
7C - 30ci F-head single, 26in wheels, magneto
7D - 50ci (820cc) F-head V-twin, 28in (711mm) wheels, magneto
Prices
7 - $225
7A - $250
7B - $225
7C - $250
7D - $300
Production
Total 5,625 (details not available)
Colours
Renault Grey with pinstriping in red, gold, blue, dark grey or black
Specifications
7D - 50ci F-head V-twin, 3 x $3\frac{1}{2}$in (76 x 89mm) bore x stroke, mechanical inlet valve, drive belt tensioner

1912

The Range
8 - 30ci (**500cc**) F-head single, battery ignition
X8 - 30ci F-head single, battery ignition, rear wheel clutch
8A - 30ci F-head single, magneto
X8A - 30ci F-head single, magneto, rear wheel clutch
8D - 50ci (**820c**) F-head V-twin, magneto
X8D - 50ci F-head V-twin, magneto, rear wheel clutch
X8E - 61ci (1000cc) F-head V-twin, magneto, rear wheel clutch, chain drive
Prices
8 - $200
X8 - $210

8A - $225
X8A - $235
8D - $275
X8D - $285
X8E - $285
Production
Total 3,852 (details not available)
Colours
As 1911
Specifications
8-X8D - Mostly as 1911, but with free-wheel clutch in rear hub of 'X'-models, operated by handle on fuel tank. Also new frame and Troxel saddle suspension system on all bikes.
X8E - 61ci flathead V-twin, $3\frac{5}{16}$in x $3\frac{1}{2}$in (84 x 89mm) bore x stroke, mechanical inlet valve, lubrication by hand-operated oil pump (replaces drip-feed as main system), free-wheel clutch, chain final drive.

1913

The Range
9A - 35ci (**575cc**) F-head single, belt drive
9B - 35ci F-head single, chain drive
9E - 61ci (**1000cc**) F-head V-twin, chain drive
9F - 61ci F-head V-twin, chain drive
9G - 61ci F-head V-twin Forecar Delivery Van
Prices
9A - $290
9B - $290
9E - $350
9F - n/a
9G - n/a
Production
9A - 1,510
9B - 4,601
9E - 6,732
9F - 49
9G - 63
TOTAL - 12,955
Colours
As 1911

TOP and ABOVE: There was now chain drive for the single, in 1913 (top) and 1914 (above) form. Harley-Davidson built over 6,000 singles in 1914, but little more than half that the following year.

Specifications
9A/B - 35ci F-head single, $3\frac{5}{16}$in x 4in (84 x 102mm) bore x stroke, mechanical inlet valve, 5hp (hence also known as 5-35)
9G - 61ci F-head V-twin with front-mounted delivery box

1914

The Range
10A - 35ci (**575cc**) F-head single, belt drive
10B - 35ci F-head single, chain drive
10C - 35ci F-head single, two-speed
10E - 61ci (**1000cc**) F-head V-twin, chain drive
10F - 61ci F-head V-twin, two-speed
10G - Delivery Van, two-speed
Prices
10A - $200
10B - $210
10C - $245
10E - $250
10F - $285
10G - n/a
Production
10A - 316
10B - 2,034
10C - 877
10E - 5,055
10F - 7,956
10G - 171
TOTAL - 16,409
Colours
As 1911
Specification
10A/B/C - Kick starter, two-speed rear hub (C), H-D's own brakes and hubs (B/C)
10E/F - Kick starter, two-speed rear hub (F), H-D's own brakes and hubs

1915

The Range
11B - 35ci (**575cc**) F-head single
11C - 35ci F-head single, two-speed
11E - 61ci (**1000cc**) F-head V-twin, single speed

11F - 61ci F-head V-twin, three-speed
11G - Delivery Van
11H - 61ci F-head V-twin, single-speed, electric lighting
11J - 61ci F-head V-twin, three-speed, electric lighting
11K - Racer, 61ci F-head V-twin

Prices
11B - $200
11C - $230
11E - $240
11F - $275
11H - $275
11J - $310
11K - $250

Production
11B - 670
11C - 545
11E - 1,275
11F - 9,855
11G - 98
11H - 140
11J - 3,719
11K - 188
TOTAL - 16,490

Colours
As 1911

Specifications
Singles - Automatic oil pump
Twins - Three-speed gearbox (11F/J), automatic oil pump, full electrical system (generator, lights, horn, ignition (11H/J)

1916

The Range
16B - 35ci (**575cc**) F-head single
16C Solo - 35ci F-head single, three-speed
16E - 61ci (**1000cc**) F-head V-twin, single-speed
16F - 61ci F-head V-twin, three-speed
16J - 61ci F-head V-twin, three-speed, electrics
16S Stripped Stock Racer - 35ci F-head single
16R - Roadster Racer - 61ci F-head V-twin
16T Track Racer - 61ci F-head V-twin

Prices
B - $200
C - $230
E - $240
F - $265
J - $295
Production
B - 292
C - 862
E - 252
F - 9,496
J - 5,898
R/S/T - 123
Delivery van - 7
TOTAL - 16,930
Colours
Renault Grey with red, black, dark blue or gold pinstriping
Specification
Kickstart (C), common frame, footboards and brake rod on all bikes

1917

The Range
17B - 35ci (**575cc**) F-head single
17C - 35ci F-head single, three-speed
17E - 61ci (**1000cc**) F-head V-twin, single-speed
17F - 61ci F-head V-twin, three-speed
17J - 61ci F-head V-twin, three-speed, electrics
17S Stripped Stock Racer - 35ci F-head single
17R Roadster Racer - 61ci F-headV-twin
17T Stripped Stock Racer - 61ci F-head V-twin
Prices
B - $215
C - $240
E - $255
F - $275
J - $310
S -
R - $280
T - $280

OPPOSITE: The 11F 61-ci (1000-cc) twin of 1915, with sidecar (a common fitment at the time). 'F' denotes a 3-speed gearbox.

ABOVE: An original unrestored 11F. Note the pushrods for the overhead inlet valves. The exhaust valves are side-mounted.

Production
B - 124
C - 605
E - 68
F - 8,527
J - 9,180
RST - 18
TOTAL - 18,522
Colours
Olive Green with Coach Green and gold pinstriping
Specifications
Singles - Dixie magneto
Twins - Four-lobe cam, revised valve timing, Dixie magneto

1918

The Range
18B - 35ci (**575cc**) F-head single, magneto
18C Solo - 35ci F-head single, magneto, three-speed
18E - 61ci (**1000cc**) F-head V-twin, single-speed, magneto
18F Solo - 61ci F-head V-twin, three-speed, magneto
18J Solo - 61ci F-head V-twin, three-speed, electrics
18R Roadster Racer - 61ci F-head V-twin, three-speed
18FUS - 61ci F-head V-twin, Presto-lite equipped, supplied to Govt. only
Prices
B - $235
C - $260
E - $275
F - $290
J - $320
Production
B - 19
C - 251
E - 5
F - 11,764
J - 6,571
R - 3

FUS - 8,095
TOTAL - 26,708
Colours
Olive Green, Coach Green, Pea Green and black pinstriping
Specifications
Singles - Berling magneto
Twins - Remy generator, many minor changes

1919

The Range
19WF Sport Twin - 35ci (**575cc**) ho twin, three-speed, magneto
19F - 61ci (**1000cc**) F-head V-twin, three-speed, magneto
19FS - 61ci F-head V-twin 'sidecar motor', three-speed
19J - 61ci F-head V-twin, three-speed, electrics
19JS - 61ci F-head V-twin 'sidecar motor', three-speed, electrics
Prices
WF - $335
F - $350
FS - $350
J - $370
JS $370
Bicycles - $30–$45
Production
W - 753
F - 5,064
J - 9,941
FS - 7,521
TOTAL - 23,279
Colours
As 1918
Specifications
W Sport Twin - 35ci horizontally-opposed twin, 2¾in x 3in (70 x 76mm) bore x stroke, 3.75:1 compression, roller-bearing crank 6hp, unit-construction three-speed gearbox, enclosed drive chain
FJ Twins - 2½in (6.4cm) wider handlebars

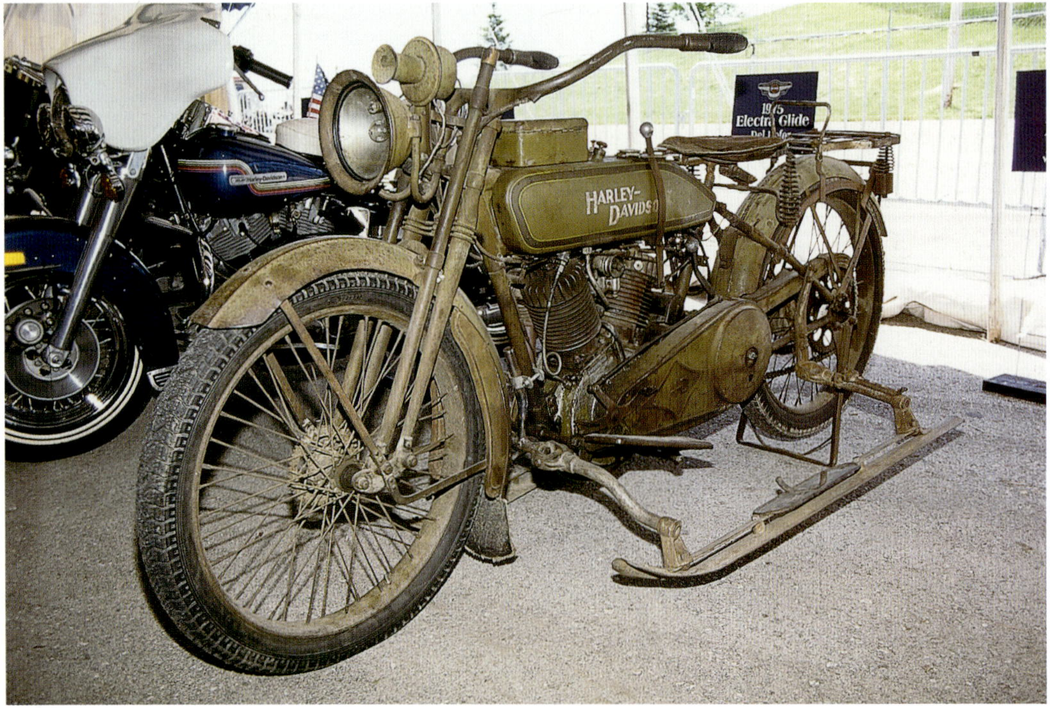

There was electric lighting for this 1919 twin, with unusual ski attachment for bad weather. There were no fair-weather bikers then!

1920

The Range
20WF Sport Twin - 35ci (**575cc**) ho twin, three-speed, magneto
20WJ Sport Twin - 35ci ho twin, three-speed, electrics
20F - 61ci (**1000cc**) F-head V-twin, three-speed, magneto
20FS - 61ci F-head V-twin, three-speed, magneto, sidecar gearing
20J - 61ci F-head V-twin, three-speed, electrics
20JS - 61ci F-head V-twin, three-speed, electrics, sidecar gearing
Prices
WF - $335
WJ - n/a
F - $370
FS - $370
J - $395
JS - $395
Production
WF - 4,459
WJ - 810
F - 7,579
J - 14,192
TOTAL - 27,040
Colours
As 1918
Specifications
WJ - Full electrics, with H-D's own generators and coils, plus lights and horn

1921

The Range

21CD Commercial - 37ci (606cc) F-head singel, three-speed, magneto
21WF - 35ci (**575cc**) ho twin, three-speed, magneto
21WJ - 35ci ho twin, three-speed, electrics
21F - 61ci (**1000cc**) F-head V-twin, three-speed, magneto
21FS - 61ci F-head V-twin, three-speed, magneto, sidecar gearing
21J - 61ci F-head V-twin, three-speed, electrics
21JS - 61ci F-head V-twin, three-speed, electrics, sidecar gearing
21FD - 74ci (**1200cc**) F-head V-twin, three-speed, magneto
21FDS - 74ci F-head V-twin, three-speed, magneto, sidecar gearing
21JD - 74ci F-head V-twin, three-speed, electrics
21JDS - 74ci F-head V-twin, three-speed, electrics, sidecar gearing

Prices

CD - $430
WF - $415
WJ - $445
F - $50
FS - $450
J - $485
JS - $485
FD - $485
FDS - $485
JD - $520
JDS - $520

Production

WF - 1,100
WJ - 823
F - 2,413
FD - 277
J - 4,526
JD - 2,321
TOTAL - 11,460

Colours

As 1918

Specifications

CD - New single for commercial use, using one cylinder of the 74ci V-twin, chassis as on V-twin
FD/JD - 74ci F-head V-twin, bored and stroked ($3^7/_{16}$in x 4in/87 x 102mm) version of 61ci, $1^1/_4$in (32mm) Schebler carburettor, 18hp.

1922

The Range

22CD Commercial - 37ci (606cc) F-head single, three-speed
22WF Sport Twin - 35ci (**575cc**) ho twin, three-speed, magneto
22WJ Sport Twin - 35ci ho twin, three-speed, electric s
22F - 61ci (**1000cc**) F-head V-twin, three-speed, magneto
22FS - 61ci F-head V-twin, three-speed, magneto, sidecar gearing
22J - 61ci F-head V-twin, three-speed electrics
22JS - 61ci F-head V-twin, three-speed, electrics, sidecar gearing
22FD - 74ci (**1200cc**) F-head V-twin, three-speed, magneto
22FDS - 74ci F-head V-twin, three-speed, magneto, sidecar gearing
22JD - F-head V-twin, three-speed, electrics
22JDS - F-head V-twin, three-speed, electrics, sidecar gearing

Prices

CD - $315
WF - $310
WJ -$315
F - $335
FD - $360
FS - $335
FDS - $360
J - $365
JS - $365
JD - $390

JDS - $390
Bicycles - $35-$45

Production

CD - 39
WF - 38
WJ - 455
F - 1,824
J - 3,183
FD - 909
JD - 3,988
TOTAL - 10,436

Colours

Brewster Green with gold pinstriping

Specifications

All big Twins - Roller-bearings for male con-rod, longer-life valve springs, double-plunger oil pump

In 1922, the FD could be yours for $360 (without the sidecar). Or there was a Harley-Davidson-badged bicycle for $35.

1923

The Range

23WF Sport Twin - 35ci (**575cc**) ho twin, three-speed, magneto
23WJ Sport Twin - 35ci ho twin, three-speed, electrics
23F - 61ci (**1000cc**) F-head V-twin, three-speed, magneto
23FS - 61ci F-head V-twin, three-speed, magneto, sidecar gearing
23J - 61ci F-head V-twin, three-speed, electrics
23JS - 61ci F-head V-twin, three-speed, electrics, sidecar gearing
23FD - 74ci (**1200cc**) F-head V-twin, three-speed, magneto
23FDS - 74ci F-head V-twin, three-speed, magneto, sidecar gearing
23JD - 74ci F-head V-twin, three-speed, electrics
23JDS - 74ci F-head V-twin, three-speed, electrics, sidecar gearing

Prices

WF - $275
WJ - $295
F - $285
FD - $310
FS - $285
FDS - $310
J - $305
JS - $305
JD - $330
JDS - $330

Production

WF - 614
WJ - 481
F - 2,822
FD - 869
J - 4,802
JD - 7,458
TOTAL - 17,046

Colours

As 1922

Specifications

FD/JD - Improved cylinder-head cooling, larger generator (JD/JDS)

1924

The Range

24FE - 61ci (**1000cc**) F-head V-twin, alloy pistons, three-speed, magneto
24FES - 61ci F-head V-twin, alloy pistons, three-speed, magneto, sidecar gearing
24JE - 61ci F-head V-twin, alloy pistons, three-speed, electrics
24JES - 61ci F-head V-twin, alloy pistons, three-speed, electrics, sidecar gearing
24FD - 74ci (**1200cc**) F-head V-twin, cast-iron pistons, three-speed, magneto
24FDS - 74ci F-head V-twin, cast-iron pistons, three-speed, magneto, sidecar gearing
24JD - 74ci F-head V-twin, cast-iron pistons, three-speed, electrics
24JDS - 74ci F-head V-twin, cast-iron pistons, three-speed, electrics, sidecar gearing
24FDCA - 74ci F-head V-twin,alloy pistons, three-speed, magneto
24FDSCA - 74ci F-head V-twin, alloy pistons, three-speed, magneto, sidecar gearing
24JDCA - 74ci F-head V-twin, alloy pistons, three-speed, electrics
24JDSCA - 74ci F-head V-twin, alloy pistons, three-speed, electrics, sidecar gearing

Prices

FE - $300
FES - $300
JE - $320
JES - $320
FD - $315
FDS- $315
JD - $335
JDS - $335
FDCA - $325
FDSCA - $325
JDCA - $345
JDSCA -$345

Production

FE - 2,708
FD - 502

FDCB - 90
FDCA - 351
JE - 4,994
JD - 2,955
JDCB - 3,034
JDCA - 3,014
TOTAL - 17,648
Colours
Olive Green, with maroon, black and gold pinstriping
Specifications
FE/JE - Aluminium-alloy pistons with four rings, Alemite grease fittings
FD/JD twins, Revised flywheels with different balance factor, Alemite grease fittings

1925

The Range
25FE - 61ci (**1000cc**) F-head V-twin, iron-alloy pistons, three-speed, magneto
25FES - 61ci F-head V-twin, iron-alloy pistons, three-speed, magneto, sidecar gearing
25JE - 61ci F-head V-twin, iron-alloy pistons, three-speed, electrics
25JES - 61ci F-head V-twin, iron-alloy pistons, three-speed, electrics, sidecar gearing
25FDCB - 74ci (**1200cc**) F-head V-twin, iron-alloy pistons, three-speed, magneto
25FDCBS - 74ci F-head V-twin, iron-alloy pistons, three-speed, magneto, sidecar gearing
25JDCB - 74ci F-head V-twin, iron-alloy pistons, three-speed, electrics
25JDCBS - F-head V-twin, iron-alloy pistons, three-speed, electrics, sidecar gearing
Prices
FE - $295
FES - $295
JE - $315
JES - $315
FDCB - $315

FDCBS- $315
JDCB - $335
JDCBS - $335
Production
F - 1,318
FD - 433
J - 4,114
JD - 9,506
TOTAL - 15,371
Colours
As 1924
Specifications
All Twins - Wider frame (lower saddle 3in/76mm), iron-alloy pistons with narrower and deeper rings, 20in (508mm) wheels, many detail changes

OPPOSITE
LEFT: The V-twin racer at the height of the board-track craze.

RIGHT: One of Harley's first purpose-built racers, the eight-valve twin (1923).

LEFT: A 1925 J-model, with the electric equipment that most buyers now preferred. This bike also has the optional speedometer.

1926

The Range

26A - 21ci (**350cc**) flathead single, three-speed, magento
26B - 21ci flathead single, three-speed, electrics
26AA - 21ci OHV single, three-speed, magento
26BA - 21ci OHV single, three-speed, electrics
26S Racer - 21ci OHV single, magneto
26J - 61ci (**1000cc**) F-head V-twin, three-speed, electrics
26JS - 61ci F-head V-twin, three-speed, electrics, sidecar gearing
26JD - 74ci (**1200cc**) F-head V-twin, three-speed, electrics
26JDS - 74ci F-head V-twin, three-speed, electrics, sidecar gearing

Prices

A - $210
B- $235
AA - $250
BA - $275
S - $300
J - $315
JS - $315
JD - $335
JDS - $335

Production

A - 1,128
AA - 207
B- 5,979
BA - 676
F - 760
FD - 232
J - 3,749
JD - 9,544
TOTAL - 22,275

Colours

As 1924

Specifications

A/B singles - 21ci $2^{7}/_{8}$in x $3^{1}/_{4}$in (73 x 83mm) bore x stroke, flat-head (A/B) with 8hp or OHV (AA/BA) with 12hp, Ricardo cylinder-heads, 26in (660mm) wheels, three-speed gearbox
J/JD - 26 x $3^{1}/_{4}$in (660 x 84mm) balloon tyres, higher output generator, bigger battery

TOP and ABOVE: Two different versions of the 1927 21-ci (350-cc) single: the utilitarian side-valve B (top) and the ohv Peashooter (above).

1927

The Range

27A - 21ci (**350cc**) flathead single, three-speed, magento
27B - 21ci flathead single, three-speed, electrics
27AA - 21ci OHV single, three-speed, magento
27BA - 21ci OHV single, three-speed, electrics
27S Racer - 21ci OHV single, magneto
27J - 61ci (**1000cc**) F-head V-twin, three-speed, electrics
27JS - 61ci F-head V-twin, three-speed, electrics, sidecar gearing
27JD - 74ci (**1200cc**) F-head V-twin, three-speed, electrics
26JDS - 74ci F-head V-twin, three-speed, electrics, sidecar gearing

Prices

A - $210
B- $235
AA - $250
BA - $275
S - $300
J - $310
JS - $310
JD - $320
JDS - $320

Production

A - 444
AA - 98
B- 3,711
BA - 524
F - 246
FD - 209
J - 3,561
JD - 9,691
TOTAL - 18,484

Colours
As 1924
Specifications
A/B Singles - Reinforced frame, new Ricardo cylinder-heads

1928

The Range
28B - 21ci (**350cc**) flathead single, three-speed, electrics
28BA - 21ci OHV single, three-speed, electrics
28J - 61ci (**1000cc**) F-head V-twin, three-speed, electrics
28JS - 61ci F-head V-twin, three-speed, electrics, sidecar gearing
28JX Sport - 61ci F-head V-twin, three-speed, electrics
28JXL Special Sport - 61ci F-head V-twin, Dow-metal pistons, three-speed, electrics
28JH Two Cam - 61ci F-head V-twin, Dow-metal pistons, three-speed, electrics
28JD - 74ci (**1200cc**) F-head V-twin, three-speed, electrics
28JDS - 74ci F-head V-twin, three-speed, electrics, sidecar gearing
28JDX Sport - 74ci F-he74 V-twin, three-speed, electrics
28JDXL Special Sport - 61ci F-head V-twin, Dow-metal pistons, three-speed, electrics
28JDH Two Cam - 74ci F-head V-twin, Dow-metal pistons, three-speed, electrics
Prices
B - $235
BA - $255
J - $310
JS - $310
JX - $310
JXL - $325
JD - $320
JDS - $320
JDX - $320
JDXL - $335
JH - $360

JDH - $390
Production
A - 519
AA - 69
B - 3,483
BA - 639
F - 141
FD - 131
J - 4,184
JD - 11,007
TOTAL - 20,173
Colours
Singles - As 1924
Twins - Main colour as 1924, but options are Black, Fawn Grey, Azure Blue, Maroon, Coach Green, White, Cream or Police Blue
Specifications
B/BA Singles - Dow-metal pistons, lighter flywheels and con-rods, revised gearing
J/JD Twins - 25in (635mm) wheels, front-wheel brake
JX/JXL/JDX/JDXL - Sports version of J/JD, with Two Cam slimmer chassis and tuned J/JD twin (bigger valves, Dow-metal pistons)
JH/JDH Two Cam Twins - Sports version of existing J/JD twins, same capacity but with direct-action valve gear, Dow-metal domed pistons, narrower fuel tank, 18in (457mm) wheels, throttle-controlled oil pump. lower, slimmer profile

ABOVE: The almost legendary JDH Two Cam of 1928.

ABOVE RIGHT: 1929 was the first year for the Forty-Five.

RIGHT: All bikes had twin headlamps for 1929.

1929

The Range

29B - 21ci (**350cc**) flathead single, three-speed, electrics
29BA - 21ci OHV single, three-speed, electrics
29C - 30.5ci (**500cc**) flathead single, three-speed, electrics
29D - 45ci (**750cc**) flathead V-twin, three-speed, electrics
29DL - 45ci flathead V-twin, high compression, three-speed, electrics
29J - 61ci (**1000cc**) F-head V-twin, three-speed, electrics
29JS - 61ci F-head V-twin, three-speed, electrics, sidecar gearing
29JH Two Cam - 61ci F-head V-twin, Dow-metal pistons, three-speed, electrics
29JD - 74ci (**1200cc**) F-head V-twin, three-speed, electrics
29JDS - 74ci F-head V-twin, three-speed, electrics, sidecar gearing
29JDH Two Cam - 74ci F-head V-twin, Dow-metal pistons, three-speed, electrics

Prices

B- $235
BA - $255
C - $255
D - $290
DL - $290
J - $310
JS - $310
JD - $320
JDS - $320
JDH - $370

Production

A - 197
AA - 26
B- 1,592
BA - 404
C - 1,570
D - 4,513
DL - 2,343
F - 191

LEFT: By 1930, Harley-Davidson was well into its side-valve gear. This one has a four-tube muffler and twin headlights.

FD - 73
J - 2,886
JD - 10,182
TOTAL - 23,977

Colours

As 1928

Specifications

C Single - 30.5ci flathead single, $3\frac{3}{32}$in x 4in (79 x 102mm) bore x stroke, 10hp, used Model B's chassis
D Twins - 45ci (750cc) flat-head V-twin, $2\frac{3}{4}$in x $3\frac{13}{16}$in (70 x 97mm) bore x stroke, Dow-metal pistons, Ricardo cylinder-heads, battery/coil ignition, three-speed gearbox

1930

The Range

30B - 21ci (350cc) flathead single, three-speed
30C - 30.5ci (500cc) flathead single, three-speed
30D - 45ci (**750cc**) flathead V-twin, low-compression, three-speed

OPPOSITE
ABOVE: A 1930 VL, a high-compression 74ci (1200cc) side-valve with sidecar.

BELOW: The 30.5-ci (500-cc) single of 1931 was not a great success.

30DS - 45ci flathead V-twin, low-compression, three-speed, sidecar model
30DL Sport - 45ci flathead V-twin, Dow-metal pistons, three-speed
30DLD Special Sport - 45ci flathead V-twin, high compression, three-speed
30V - 74ci (**1200cc**) flathead V-twin, medium-compression, three-speed
30VL Sport - 74ci flathead V-twin, high-compression, Dow-metal pistons
30VS - 74ci flathead V-twin, Dow-metal pistons, three-speed, sidecar model
30VC Commercial - 74ci flathead V-twin, nickel-iron pistons, three-speed

Prices
B- $235
BA -$255
C -$260
D -$310
DS -$310
DL - $310
DLD -$310
V (all models) - $340

Production
B - 577
BA - 767
C - 1,627
D - 2,000
DS - 213
DL -3,191
DLD - 206
V - 1,960
VS - 3,612
VL -3,246
VC - 1,174
TOTAL - 18,573

Colours
Singles - Olive Green with vermilion, maroon and gold pinstriping
Twins - As singles, with options of Maroon, Cream, Azure Blue, Fawn Grey, Coach Green or Police Blue. Any combination of two colours could be ordered as well.

Specifications
C - Now uses Model D's new frame (lower saddle, more ground clearance), I-beam front forks, automatic primary chain oiling
D - New frame (lower saddle, more ground clearance), I-beam front forks, auto primary chain oiling
V - New flat-head (side-valve) Big Twin to replace the J Model. 74ci flathead V-twin, 3$^{7}/_{16}$in x 4in (87 x 102mm) bore x stroke, Ricardo cylinder heads, magnesium-alloy pistons, total loss lubrication, three-speed gearbox, dual headlights (1930 only)

1931

The Range
31B - 21ci (350cc) flathead single - now for export only
31C - 30.5ci (500cc) flathead single, three-speed
31D - 45ci (**750cc**) flathead V-twin, low-compression, three-speed
31DS - 45ci flathead V-twin, low-compression, three-speed, sidecar gearing
31DL Sport - 45ci flathead V-twin, high-compression, three-speed
31DLD Special Sport - 45ci flathead V-twin, high-compression, three-speed
31V - 74ci (**1200cc**) flathead V-twin, medium-compression, three-speed
31VL Sport - 74ci flathead V-twin, high-compression, Dow-metal pistons
31VS - 74ci flathead V-twin, medium-compression, three-speed, sidecar gearing
31VC Commercial - 74ci flathead V-twin, low-compression, three-speed

Prices
C - $260
D - $310
DS - $310
DL - $310
DLD - $325
V (all models) - $340

Production
C - 1,059
D - 715

DS - 276
DL - 1,306
DLD - 241
V - 825
VS - 1,994
VL - 3,477
TOTAL - 9,893
Colours
Olive Green with vermilion, maroon and gold pinstriping, with options of White, Maroon/Cream, Black/Vermilion, Olive Green/Vermilion, or Police Blue/Fawn Grey.
Specifications
D Twins - Reinforced frame (as per Tom Sifton's recommendation), Schebler De Luxe carburettor, new generator drive gears, new clutch assembly

1932

The Range
32B - 21ci (350cc) flathead single, three-speed
32C - 30.5ci (500cc) flathead single, three-speed
32R - 45ci (**750cc**) flathead V-twin, low-compression, three-speed
32RS - 45ci flathead V-twin, low-compression, three-speed, sidecar model
32RL Sport - 45ci flathead V-twin, Dow-metal pistons, three-speed
32RLD Special Sport - 45ci flathead V-twin, high-compression, three-speed
32V - 74ci (**1200cc**) flathead V-twin, medium-compression, three-speed
32VL Sport - 74ci flathead V-twin, high-compression, Dow-metal pistons

ABOVE and TOP: Both these 1932/33 VLs show the art deco bird motif, used on all models in 1933.

32VS - 74ci flathead V-twin, medium-compression, three-speed, sidecar gearing
32VC Commercial - 74ci flathead V-twin, low-compression, three-speed
32G Servicar - 45ci flathead V-win, three-speed
32GD Servicar - 45ci flathead V-twin, three-speed, large compartment
32GE Servicar - 45ci flathead V-twin, three-speed, air tank
Prices
B - $195
C - $235
R - $295
RS - $295
RL - $295
RLD - $310
V (all models) - $320
G Servicar - $450
Production
B - 535
C - 540
R - 410
RS - 111
DL - 628
DLD - 98
V - 478
VS - 1,233
VL - 2,684
VC - 239
G/GA Servicar - 219
TOTAL - 7,175
Colours
Main colour as 1931, with options of Delft Blue/Turquoise, Vermilion, Vermilion/Black, White, Police Blue or Delft Blue
Specifications
R Twins - Replaces D twins. 45ci flathead V-twin with stronger valve springs, more air space between cylinders, larger flywheels, new crankcases, aluminium pistons, heavy-duty oil pump, Linkert carburettor

V Twins - More air space between cylinders, improved oil pump, new generator (more consistent current, better cooling), heavy-duty front forks
G Servicars - Basically a three-wheeler version of the Model R, for commercial use, single drum brake inside rear axle casing

1933

The Range

33B - 21ci (350cc) flathead single, three-speed
33CB - 30.5ci (**500cc**) flathead single, three-speed
33C - 30.5ci flathead single, three-speed
33R - 45ci (**750cc**) flathead V-twin, low-compression, three-speed
33RS - 45ci flathead V-twin, low-compression, three-speed, sidecar model
33RL Sport - 45ci flathead V-twin, high compression, three-speed
33RLD Special Sport - 45ci flathead V-twin, high-compression, three-speed
33V - 74ci (**1200cc**) flathead V-twin, medium-compression, three-speed
33VL Sport - 74ci flathead V-twin, high-compression, three-speed
33VS - 74ci flathead V-twin, medium-compression, three-speed, sidecar gearing
33VLD - 74ci flathead V-twin, high compression, magnesium-alloy pistons, three-speed
33VC Commercial - 74ci flathead V-twin, low-compression, three-speed
33G Servicar - 45ci flathead V-win, three-speed+reverse
33GD Servicar - 45ci flathead V-twin, three-speed, large compartment
33GE Servicar - 45ci flathead V-twin, three-speed, air tank

Prices

B - $187.50
C - $225
R - $280
RS - $280
RL - $280
RLD - $290
V (all except VLD) - $310
VLD - $320

Production

B - 123
C - 183
R - 162
RS - 3
RL - 264
RLD - 68
V - 233
VS - 164
VL - 886
VLD - 780
VC - 109
UFS - 499
G/GA Servicar - 182
TOTAL - 3,656

Colours

Singles & Servicars - Silver/Turquoise with Black/Gold strips and bird motif
CB Single - Olive Green/Black with bird motif
Twins (except Servicar) - Sunshine Blue/White with gold striping, Silver/Turquoise, Olive/Brilliant Green, Black/Mandarin Red or Police Blue/White, all with bird motif.

Specifications

CB - New model, using 30.5ci single in Model B (21ci) frame).
Otherwise, no major changes

1934

The Range

34B - 21ci (350cc) flathead single, three-speed
34CB - 30.5ci (**500cc**) flathead single, three-speed
34C - 30.5ci flathead single, three-speed
34R - 45ci (**750**cc) flathead V-twin, low-compression, three-speed
34RS - 45ci flathead V-twin, low-compression, three-speed, sidecar model
34RL Sport - 45ci flathead V-twin, three-speed
34RLD Special Sport - 45ci flathead V-twin, high-compression, three-speed
34VD - 74ci (**1200cc**) flathead V-twin, low-compression, three-speed
34VDS - 74ci flathead V-twin, low-compression, three-speed, sidecar gearing
34VFDS - 74ci flathead V-twin, low-compression, nickel-iron pistons, three-speed, sidecar gearing
34VLD Special Sport - 74ci flathead V-twin, high compression, three-speed+reverse
34G Servicar - 45ci flathead V-win, three-speed+reverse
34GD Servicar - 45ci flathead V-twin, three-speed+reverse, large compartment
34GE Servicar - 45ci flathead V-twin, three-speed, air tank

1934, and the tank loses that beautiful art deco bird in favour of a winged diamond. Harley-Davidson was weathering the Depression, with very low sales, and closure was on the agenda. To save money, plain black details replaced chrome or nickel plate.

Prices
B - $187.50
C - $225
CB - $197.50
R - $280
RL - $280
RLD - $290
V (all models) - $310
G Servicar - $430
GD Servicar - $430
GE Servicar - $485
Production
B - 424
C - 220
CB - 310
R - 450
RS - 302
RL - 743
RLD - 240
VD - 664
VDS - 1,029
VL - 886
VLD - 4,527
VFD/VFDS - 1,335
Servicars - 229
TOTAL - 11,359
Colours
Singles - Silver/Red Teak or Olive Green/Black with Art Deco logo
CB Single - As 1933, with Art Deco logo
Twins - Silver/Red Teak, Red Teak/Black, Black/Orlando Orange, Seafoam Blue/Silver, Black/Olive Green, Vermilion/Copper, all with Art Deco logo
Specifications
R Twins - Aluminium-alloy pistons, new oil pump, new clutch
V Twins - Aluminium-alloy pistons, High-flo muffler
Servicars - Straight-bore cylinders with T-slot pistons

$310 would buy you a brand new VL in 1934, but few could afford it.

In 1935, chromium began to make a tentative appearance on handlebars and headlamps, but it would be several years before sales fully recovered to pre-Depression levels.

1935

The Range

35R - 45ci (**750cc**) flathead V-twin, low-compression, three-speed
35RS - 45ci flathead V-twin, low-compression, three-speed, sidecar model
35RL - 45ci flathead V-twin, high-compression, three-speed
35RLD Special Sport - 45ci flathead V-twin, three-speed
35RLDR Competition Special - 45ci flathead V-twin, three-speed
35VD - 74ci (**1200cc**) flathead V-twin, low-compression, three-speed
35VDS - 74ci flathead V-twin, low-compression, three-speed, sidecar gearing
34VFDS - 74ci flathead V-twin, low-compression, nickel-iron pistons, three-speed, sidecar gearing
35VLD Special Sport - 74ci flathead V-twin, high compression, three-speed
35VLDJ Competition Special - 74ci flathead V-twin, three-speed
35VLDD Sport Solo - 80ci (**1300cc**) flathead V-twin, high compression, three-speed, solo bars
35VDDS - 80ci flathead V-twin, three-speed, sidecar model
35G Servicar - 45ci flathead V-win, three-speed+reverse
35GD Servicar - 45ci flathead V-twin, three-speed+reverse, large compartment
35GE Servicar - 45ci flathead V-twin, three-speed+reverse, air tank

Prices

R - $295
RL - $295
RS - $295
RLD - $305
RLDR - $322
VD - $320
VLD - $320
VDS - $320
VLDJ - $333.50
VLDD - $347
VDDS - $347
G Servicar - $440
GD Servicar - $440
GE Servicar - $495

Production

R - 543
RS - 392
RL - 819
RLD - 177
RLDR - 29
VD - 585
VDS - 1,189
VL - 886
VLD - 3,963
VLDJ - 102
VFDS - 327
VLDD -179
Servicars - 567
TOTAL - 9,758

Colours

Egyptian Ivory/Regent Brown, Teal Red/Black, Olive Green/Black, Verdant Green/Black, Venetian Blue/Silver, Silver/Black (police only), Cameo Cream/Dawn Grey, Potomac Blue

Specifications

All models - Carburized braked drums and harder linings
R Twins - Constant-mesh gearbox, quickly-detachable rear wheel
V Twins - New 80ci flathead V-twin, 3.422 x 3.125in (87 x 79mm) bore x stroke, same chassis as 74ci. Four-speed transmission option for sidecar use

1936

The Range

36R - 45ci (**750cc**) flathead V-twin, low-compression, three-speed
36RS - 45ci flathead V-twin, low-compression, three-speed, sidecar gearing
36RL - 45ci flathead V-twin, high-compression, three-speed
36RLD Special Sport - 45ci flathead V-twin, high compression, three-speed
36RLDR Competition Special - 45ci flathead V-twin, three-speed
36E - 61ci (**1000cc**) OHV V-twin, medium-compression, four-speed
36EL Special Sport - 61ci OHV V-twin, high-compression, four-speed
36ES - 61ci OHV V-twin, medium-compression, four-speed, sidecar gearing
36VD - 74ci (**1200cc**) flathead V-twin, low-compression, three-speed
36VDS - 74ci flathead V-twin, low-compression, three-speed, sidecar gearing
36VLH Sport - 80ci (**1300cc**) flathead V-twin, high compression, three-speed, solo bars
36VHS - 80ci flathead V-twin, three-speed, sidecar gearing
36G Servicar - 45ci flathead V-win, three-speed+reverse
36GD Servicar - 45ci flathead V-twin, three-speed+reverse, large compartment
36GE Servicar - 45ci flathead V-twin, three-speed+reverse, air tank

Prices

R - $295
RL - $295
RS - $295

RLD - $305
RLDR -$320
E -$380
EL - $380
ES - $380
VD - $320
VLD - $320
VDS - $320
VLH - $340
VHS - $340
G Servicar - $440
GD Servicar - $440
GE Servicar - $495
Production
R - 539
RS - 437
RL - 355
RLD - 540
RLDR - 79
E - 152
EL - 1,526
ES - 26
VD - 176
VDS - 623
VL - 886
VLD - 1,577
VLH - 2,046
VHS - 305
VFHS - 35
VFDS - 600
Servicars - 648
TOTAL - 10,550
Colours
Sherwood Green/Silver, Dusk Grey/Royal Buff, Teak Red/Black, Venetian Blue/Croydon Cream, Maroon/Nile Green
Specifications
R 45ci Twins - Deeper finning on heads and cylinders, redesigned combustion chambers, 1.25in (32mm) carburettor (RL/RLD), many minor changes
E 61ci Twins - All-new OHV 'Knucklehead' V-twin, 61ci, 3 5/16in x 3 1/2in 84 x 89mm) bore x stroke, dry sump lubrication, four-speed gearbox, heavy duty clutch, saddle-type twin fuel tanks, 18in wheels
V 74ci/80ci Twins - Deeper finning on heads and cylinders, new combustion chamber, four-speed gearbox optional, many minor changes
Servicars - Deeper finning on cylinders, redesigned combustion chambers, many minor changes

1936 was the first year of the 80-ci (1300-cc) side-valve twin – the biggest Harley yet.

1937

The Range
37W - 45ci (**750cc**) flathead V-twin, low-compression, three-speed
37WS - 45ci flathead V-twin, low-compression, three-speed, sidecar gearing
37WL - 45ci flathead V-twin, high-compression, three-speed
37WLD Special Sport - 45ci flathead V-twin, high compression, three-speed
37WLDR Competition Special - 45ci flathead V-twin, three-speed
37E - 61ci (**1000cc**) OHV V-twin, medium-compression, four-speed
37EL Special Sport - 61ci OHV V-twin, high-compression, four-speed
37ES - 61ci OHV V-twin, medium-compression, four-speed, sidecar gearing
37U - 74ci (**1200cc**) flathead V-twin, low-compression, four-speed
37UL Special Sport - 74ci flathead V-twin, high-compression, four-speed
37US - 74ci flathead V-twin, low-compression, four-speed, sidecar gearing
37UH - 80ci (**1300cc**) flathead V-twin, medium-compression, four-speed
37ULH Sport - 80ci flathead V-twin, high compression, four-speed
37UHS - 80ci flathead V-twin, medium-compression, four-speed, sidecar gearing
37G Servicar - 45ci flathead V-win, three-speed+reverse
37GD Servicar - 45ci flathead V-twin, three-speed+reverse, large compartment
37GE Servicar - 45ci flathead V-twin, three-speed+reverse, air tank

Prices
W - $355
WL - $355
WS - $355
WLD - $355
WLDR - $380
E - $435
EL -$435
ES - $435
U - $395
UL - $395
US - $395
UH - $415
ULH -$415
UHS - $415
G Servicar - $515
GD Servicar - $515
GE Servicar - $570
Production
W - 539
WS - 232
WL - 560
WLD - 581
WLDR - 145
E - 126
EL - 1,829
ES - 70
U - 612
US - 1,080
UL -2,861
UH - 185
ULH - 1,513
UHS - 400
Servicars - 816
TOTAL - 11,549
Colours
Teal Red/Black, Bronze Brown/Delphine Blue, Police Silver/Black, Delphine Blue/Teal Red, Olive Green/Black
Specifications
W 45ci Twins - redesigned R-series, with styling from the E-series Knucklehead and a redesigned 45ci flathead V-twin with dry sump lubrication, roller bearings, saddle fuel tank, tanktop instrument panel
E 61ci Twins - No major changes
U 74ci/80ci Twins - redesigned V-series, with styling from the E-series Knucklehead and redesigned engines with dry sump lubrication, roller bearings, E-model four-speed gearbox, frame and running gear, saddle fuel tank, tanktop instrument panel, automatic primary chain oiling
Servicars - Used styling from E-series Knucklehead and redesigned 45ci V-twin from W-series (see above)

1938

The Range
38WL - 45ci (**750cc**) flathead V-twin, high-compression, four-speed
38WLD Sport - 45ci flathead V-twin, high compression, four-speed
38WLDR Competition Special - 45ci flathead V-twin, four-speed
38E - 61ci (**1000cc**) OHV V-twin, medium-compression, four-speed
38EL Special Sport - 61ci OHV V-twin, high-compression, four-speed
38ES - 61ci OHV V-twin, medium-compression, four-speed, sidecar gearing
38U - 74ci (**1200cc**) flathead V-twin, medium-compression, four-speed
38UL Special Sport - 74ci flathead V-twin, high-compression, four-speed
38US - 74ci flathead V-twin, medium-compression, four-speed, sidecar gearing
38UH - 80ci (**1300cc**) flathead V-twin, medium-compression, four-speed
38ULH Special Sport - 80ci flathead V-twin, high compression, four-speed
38UHS - 80ci flathead V-twin, medium-compression, four-speed, sidecar gearing
38G Servicar - 45ci flathead V-win, three-speed+reverse
38GD Servicar - 45ci flathead V-twin, three-speed+reverse, large compartment
Prices
WL - $355
WLD - $355
WLDR - $380
E - $435
EL -$435
ES - $435
U - $395
UL - $395
US - $395
UH - $415
ULH - $415
UHS - $415
G Servicar - $515
GD Servicar - $515
Production
W - 302
WS - 247
WL - 309
WLD - 402
WLDR - 139
EL - 2,289
ES - 189
U - 504
US - 1,193
UL - 1,099
UH - 108
ULH - 579
UHS - 132
Servicars - 525
TOTAL - 8,017
Colours
Hollywood Green with gold striping, Teak Red/Black, Venetian Blue/White, Police Silver/Black (police), Silver Tan/Sunshine Blue, Olive Green/Black
Specifications
W 45ci Twins - Many minor changes
E 61ci Twins - Redesigned gearbox, many minor changes
U/UH 74ci/80ci Twins - Many minor changes
Servicars - Many minor changes

1938 and the Servicar gets a helping hand. The three-wheeler was to be Harley-Davidson's longest-running model.

1939

The Range

39WL - 45ci (**750cc**) flathead V-twin, high-compression, four-speed
39WLD Sport - 45ci flathead V-twin, high compression, four-speed
39WLDR Competition Special - 45ci flathead V-twin, four-speed
39EL Special Sport - 61ci (**1000cc**) OHV V-twin, high-compression, four-speed
39ES - 61ci OHV V-twin, medium-compression, four-speed, sidecar gearing
39U - 74ci (**1200cc**) flathead V-twin, medium-compression, four-speed
39UL Sport - 74ci flathead V-twin, high-compression, four-speed
39US - 74ci flathead V-twin, medium-compression, four-speed, sidecar gearing
39UH - 80ci (**1300cc**) flathead V-twin, medium-compression, four-speed
39ULH Special Sport - 80ci flathead V-twin, high compression, four-speed
39UHS - 80ci flathead V-twin, medium-compression, four-speed, sidecar gearing
39G Servicar - 45ci flathead V-win, three-speed+reverse
39GD Servicar - 45ci flathead V-twin, three-speed+reverse, large compartment

Prices

WL - $355
WLD - $355
WLDR - $380
EL - $435
ES - $435
U - $395
UL - $395
US - $395
UH - $415
ULH - $415
UHS - $415
G Servicar - $515
GD Servicar - $515

Production

W - 260
WS - 170

WL - 212
WLD - 326
WLDR -173
EL - 2,695
ES - 214
U - 421
US - 1,327
UL - 902
UH - 92
ULH - 384
UHS -109
Servicars - 525
TOTAL - 7,810

Colours

Black with Ivory panel, Airway Blue/White, Teak Red/Black, Police Silver/Black

Specifications

W 45ci Twins - Revised pistons/rings, longer-life valve springs, 'cat's-eye' streamlined instrument panel
E 61ci Twins - Revised pistons/rings, improved valve springs, optional three-speed gearbox, cat's-eye instrument panel
U 74ci/80ci Twins - Revised piston rings, longer inlet manifold (to cool carburettor), thinner cylinder walls (80ci), new four-speed gearbox, cat's-eye instrument panel
Servicars - As W 45ci Twins

1940

The Range

40WL Sport - 45ci (**750cc**) flathead V-twin, high-compression, four-speed
40WLD Special Sport - 45ci flathead V-twin, aluminium cylinder heads, high compression
40WLDR Competition - 45ci flathead V-twin, aluminium heads
40EL Special Sport - 61ci (**1000cc**) OHV V-twin, high-compression, four-speed
40ES - 61ci OHV V-twin, medium-compression, four-speed, sidecar gearing
40U - 74ci (**1200cc**) flathead V-twin, medium-compression, four-speed
40US - 74ci flathead V-twin, medium-compression, four-speed, sidecar gearing
40UL Special Sport - 74ci flathead V-twin, four-speed
40UH - 80ci (**1300cc**) flathead V-twin, medium-compression, four-speed
40ULH Special Sport - 80ci flathead V-twin, high compression, four-speed
40UHS - 80ci flathead V-twin, medium-compression, four-speed, sidecar gearing
40G Servicar - 45ci flathead V-win, three-speed+reverse
40GD Servicar - 45ci flathead V-twin, three-speed+reverse, large compartment

Prices

WL - $350
WLD - $350
WLDR -$395
EL - $430
ES - $430
U - $385
UL - $385
US - $385
UH - $410
ULH - $410
UHS - $410
G Servicar - $515
GD Servicar - $515

Production

W - 439
WS - 202
WL - 569
WLD - 567
WLDR - 87
EL - 3,893
ES - 176
U - 260
US - 1,516
UL - 822
UH - 187
ULH - 672
UHS - 163
Servicars - 908
TOTAL - 10,461

OPPOSITE: The Knucklehead (this is a 1939 example) heralded a new era of high performance for Harley-Davidson.

RIGHT: This 1940 WL45 is in civilian trim, but most of those built over the next few years would be to a military specification. It was at this time the smallest bike Harley made, and the workhorse of the range.

FAR RIGHT: A 1940 Knucklehead.

Colours
Black with Flight Red strip, Clipper Blue/White, Squadron Grey/Bittersweet, Flight Red/Black, Police Silver/Black (police only)

Specifications
W 45ci Twins - Aluminium heads, deep-finned cylinders, larger carburettor (WLD, WLDR), 5 x16in (127 x 406mm) tyres optional
E 61ci Twins - New piston rings, rear crankcase baffle removed, larger main-bearing, larger intake ports, manifold and $1^1/_2$in (38mm) carburettor, many minor changes
U 74ci/80ci Twins - Larger main-bearing, constant-mesh four-speed gearbox, aluminium cylinder heads (80ci), many minor changes
Servicars - As W 45ci Twins

1941

The Range
41WL - 45ci (**750cc**) flathead V-twin, four-speed
41WLD - Sport - 45ci flathead V-twin
41WLDR - Special Sport - 45ci flathead V-twin, aluminium heads
41WLA - 45ci flathead V-twin, US Army
41WLC - 45ci flathead V-twin, Canadian Army
41EL Special Sport - 61ci (**1000cc**) OHV V-twin, high-compression, four-speed
41ES - 61ci OHV V-twin, medium-compression, four-speed, sidecar gearing
41U - 74ci (**1200cc**) flathead V-twin, medium-compression, four-speed
41US - 74ci flathead V-twin, medium-compression, four-speed, sidecar gearing
41UL Special Sport - 74ci flathead V-twin, high-compression, four-speed
41FL Special Sport - 74ci OHV V-twin, high-compression, four-speed
41FS - 74ci OHV V-twin, medium-compression, four-speed, sidecar gearing
41UH - 80ci (**1300cc**) flathead V-twin, medium-compression, four-speed
41ULH Special Sport - 80ci flathead V-twin, high compression, four-speed
41UHS - 80ci flathead V-twin, medium-compression, four-speed, sidecar gearing
41G Servicar - 45ci flathead V-win, three-speed+reverse
41GD Servicar - 45ci flathead V-twin, three-speed+reverse, large compartment

Prices
WL - $350
WLD - $365
WLDR - $385
E - $425
EL - $425
ES - $430
FL - $465
FS - $465
U - $385
UL - $385
US - $385
UH - $410
ULH - $410
UHS - $410
G Servicar - $515
GD Servicar - $515

Production
W - 4,095
WL - 4,277
WLA - 2,282
WLC - 149

WLD - 455
WLDR -171
EL - 2,280
ES - 261
FL - 2,452
FS - 156
U - 884
US - 1,888
UL - 715
UH - 126
ULH - 420
UHS - 1,112
Servicars - 1,159
TOTAL - 22,882

Colours

Brilliant Black, Skyway Blue, Flight Red, Cruiser Green, Police Silver (police only)

Specifications

W 45ci Twins - Many minor changes
E/F 61/74ci Twins - New F-series 74ci OHV V-twin is enlarged version of 61ci, with $3\frac{7}{16}$in x $3\frac{31}{32}$in (87 x 101mm) bore x stroke, heavier flywheels. All E/F twins had redesigned crankcases, clutch, revised geometry (29-degree rake), 16in (406mm) wheels
U/UH 74/80ci Twins - Many minor changes
Servicars - As W 45ci Twins

1942

The Range

42WL - 45ci (**750cc**) flathead V-twin, four-speed
42WLD Special Sport - 45ci flathead V-twin, aluminium heads, four-speed
42WLA - 45ci flathead V-twin, US Army
42WLC - 45ci flathead V-twin, Canadian Army
42E - 61ci (**1000cc**) OHV V-twin, high-compression, four-speed
42EL Special Sport - 61ci OHV V-twin, medium-compression, four-speed
42U - 74ci (**1200cc**) flathead V-twin,

ABOVE: The 80-ci (1300-cc) side-valve twin was soon to be replaced by the 74-ci (1200-cc) Knucklehead. Even as war clouds gathered, the evolution of the range continued.

OPPOSITE: War Child: Harley-Davidson sold thousands of WLA 45s to the military, though few of them survived.

medium-compression, four-speed
42UL Special Sport - 74ci flathead V-twin, high-compression, four-speed
42F - 74ci OHV V-twin, medium-compression, four-speed
42FL Special Sport - 74ci OHV V-twin, high-compression, four-speed
42XA - 45ci flathead ho twin, US Army
42G Servicar - 45ci flathead V-win, three-speed+reverse

Prices

WL - $350
WLD - $365
E - $425
EL - $425
F -$465
FL - $465
U - $385
UL - $385
UH - $410
ULH - $410
UHS - $410
G Servicar - $525
GD Servicar - $510

Production

WLA - 13,051
WLC - 9,820
WL - 142
WLD - 133
XA - 1,011
EL - 620
ELA - 8
ELC - 45
ES - 164
FL - 799
FS - 107
U - 421
US - 978
UL - 405
Servicars - 399
TOTAL - 28,103

Colours

As 1941

Specifications

No major changes

1943

The Range

43WLA - 45ci (**750cc**) flathead V-twin, US Army
43WLC - 45ci flathead V-twin, Canadian Army
43E - 61ci (**1000cc**) OHV V-twin, medium-compression, four-speed
43EL Special Sport - 61ci OHV V-twin, high-compression, four-speed
43U - 74ci (**1200cc**) flathead V-twin, medium-compression, four-speed
43UL - 74ci flathead V-twin, high-compression, four-speed
43F - 74ci OHV V-twin, medium-compression, four-speed
43FL Special Sport - 74ci OHV V-twin, high-compression, four-speed
43XA - 45ci flathead ho twin, US Army
43G Servicar - 45ci flathead V-win, three-speed+reverse

Prices

E - $425
EL - $425
F - $465
FL - $465
U - $385
UL - $385

Production

WLA - 24,717
WLC - 2,647
EL - 53
ES - 105
FL - 33
FS - 112
U - 493
US - 1,315
UL - 11
Servicars - 135
TOTAL - 29621

Colours

Grey or Silver

Specifications

No major changes

1944

The Range

42WLA - 45ci (**750cc**) flathead V-twin, US Army
43WLC - 45ci flathead V-twin, Canadian Army
44E - 61ci (**1000cc**) OHV V-twin, medium-compression, four-speed
44EL Special Sport - 61ci OHV V-twin, high-compression, four-speed
44U - 74ci (**1200cc**) flathead V-twin, medium-compression, four-speed
44UL - 74ci flathead V-twin, high-compression, four-speed
44F - 74ci OHV V-twin, medium-compression, four-speed
44FL Special Sport - 74ci OHV V-twin, high-compression, four-speed
44G Servicar - 45ci flathead V-win, three-speed+reverse

Prices

E - $425
EL - $425
F - $465
FL - $465
U - $385
UL - $385

Production

WLA - 11,351
WLC - 5,356
WL - 57
EL - 116
ES - 105
FL - 172
FS - 67
U - 580
US - 206
UL - 1366
Servicars - 57
TOTAL - 19,433

Colours

As 1943

Specifications

W 45ci Twins - No major changes
E/F 61/74ci Twins - Synthetic rubber tyres
U 74ci Twins - Synthetic rubber tyres
Servicars - Synthetic rubber tyres

ABOVE: The FL was the enlarged 74-ci (1200-cc) version of the Knucklehead, though the 61-ci (1000-cc) EL continued alongside it.

OPPOSITE: 1946, and peacetime and chrome returns with a vengeance to this FL, on silencer, wheel rims, front hub and headlamp

1945

The Range

45WLA - 45ci (**750cc**) flathead V-twin, four-speed, US Army
45WLR - 45ci flathead V-twin, four-speed, Russian Army
45WL - 45ci flathead V-twin, four-speed
45E - 61ci (**1000cc**) OHV V-twin, medium-compression, four-speed
45EL Special Sport - 61ci OHV V-twin, high-compression, four-speed
45ES - 61ci OHV V-twin, medium-compression, four-speed, sidecar gearing
45U Commercial - 74ci (**1200cc**) flathead V-twin, medium-compression, four-speed
45UL - 74ci flathead V-twin, high-compression, four-speed
45US - 74ci flathead V-twin, medium compression, four-speed, sidecar gearing
45F - 74ci OHV V-twin, medium-compression, four-speed

45FL Special Sport - 74ci OHV V-twin, high-compression, four-speed
45FS - 74ci OHV V-twin, four-speed, sidecar gearing
45G Servicar - 45ci flathead V-win, three-speed+reverse

Prices
WL - $395.97
E - $463.67
EL - $463.67
ES - $463.67
F - $465
FL - $465
FS - $465
U - $427.25
UL - $427.25
US - $427.25
Servicar - $580.33
XA - $500

Production
WLA - 8,317
WL - 1,357
EL - 398
ES - 282
FL - 619
FS - 131
U - 513
US - 217
UL - 555
Servicars - 86
TOTAL - 12,475

Colours
Grey

Specifications
W 45ci Twins - Aluminium cylinder heads optional
E/F 61/74ci Twins - No changes
U 74ci Twins - Aluminium cylinder heads optional

1946

The Range
46WL - 45ci (750cc) flathead V-twin, three-speed

46E - 61ci (**1000cc**) OHV V-twin, medium-compression, four-speed
46EL Special Sport - 61ci OHV V-twin, high-compression, four-speed
46ES - 61ci OHV V-twin, medium-compression, four-speed, sidecar gearing
46U - 74ci (**1200cc**) flathead V-twin, medium-compression, four-speed
46UL - 74ci flathead V-twin, high-compression, four-speed
46US - 74ci flathead V-twin, medium compression, four-speed, sidecar gearing
46F - 74ci OHV V-twin, medium-compression, four-speed
46FL Special Sport - 74ci OHV V-twin, high-compression, four-speed
46FS - 74ci OHV V-twin, medium-speed, four-speed, sidecar gearing
46G Servicar - 45ci flathead V-win, three-speed+reverse

Prices
WL - $395.97
E - $463.67
EL - $463.67
ES - $463.67
F - $465
FL - $465
FS - $465
U - $427.25
UL - $427.25
US - $427.25
Servicar - $593.93

Production
WL - 4,410
EL - 2,098
ES - 244
FL - 3,986
FS - 418
U - 670
US - 1,052
UL - 1,800
Servicars - 1,444
TOTAL - 16122

Colours
Grey or Flight Red

Specifications
No major changes

1947

The Range
47WL - 45ci (750cc) flathead V-twin, three-speed
47E - 61ci (**1000cc**) OHV V-twin, medium-compression, four-speed
47EL Special Sport - 61ci OHV V-twin, high-compression, four-speed
47ES - 61ci OHV V-twin, medium-compression, four-speed, sidecar gearing
47U - 74ci (**1200cc**) flathead V-twin, medium-compression, four-speed
47UL - 74ci flathead V-twin, high-compression, four-speed
47US - 74ci flathead V-twin, medium compression, four-speed, sidecar gearing
47F - 74ci OHV V-twin, medium-compression, four-speed
47FL Special Sport - 74ci OHV V-twin, high-compression, four-speed
47FS - 74ci OHV V-twin, medium-speed, four-speed, sidecar gearing
47G Servicar - 45ci flathead V-win, three-speed+reverse

Prices
WL - $490
E - 590
EL - $590
ES - $590
F - $605
FL - $605
FS - $465
U - $545
UL - $545
US - $545
Servicar - $710

Production
WL - 3,338
EL - 4,117
ES - 237
FL - 6,893
FS - 401
U - 422
US - 1,267

UL - 1,243
Servicars - 2,177
TOTAL - 20,095
Colours
Brilliant Black, Flight Red, Skyway Blue, Police Silver (police only), Olive Green
Specifications
No major changes

1948

The Range
48S - 7ci (125cc) two-stroke single, three-speed
48WL - 45ci (750cc) flathead V-twin, three-speed
48E Sport - 61ci (**1000cc**) OHV V-twin, medium-compression, four-speed
48EL Special Sport - 61ci OHV V-twin, high-compression, four-speed
48ES - 61ci OHV V-twin, medium-compression, four-speed, sidecar gearing
48U Sport - 74ci (**1200cc**) flathead V-twin, medium-compression, four-speed
48UL Sport Special - 74ci flathead V-twin, high-compression, four-speed
48US - 74ci flathead V-twin, medium compression, four-speed, sidecar gearing
48F Sport - 74ci OHV V-twin, medium-compression, four-speed
48FL Special Sport - 74ci OHV V-twin, high-compression, four-speed
48FS - 74ci OHV V-twin, medium-speed, four-speed, sidecar gearing
48G Servicar - 45ci flathead V-win, three-speed+reverse
Prices
WL - $535
E - $635
EL - $635
ES - $635
F - $650
FL - $650
FS - $650
U - $590
UL - $590

US - $590
Servicar - $755
Production
S - 10,117
WL - 2,124
EL - 4,321
ES - 198
FL - 8,071
FS - 334
U - 401
US - 1,006
UL - 970
Servicars - 1,778
TOTAL - 29,320
Colours
Brilliant Black, Flight Red, Azure Blue, Police Silver (police only)

Specifications
S 7ci Single - 125cc two-stroke, 2 1/6in x 2 9/32in (52 x 60mm) bore x stroke, 6.6:1, 1.7hp, three-speed gearbox, rigid frame, girder forks, six-volt electrics, 3 1/4 x 19in (82.5 x 483mm) tyres
W 45ci V-twin - No major changes
E/F 61/74ci Twins - Revised 'Panhead' motor with new top end (hydraulic tappets and aluminium heads), new wishbone frame
U 74ci Twins - New wishbone frame
Servicars - No major changes

1949

The Range
49S - 7ci (125cc) two-stroke single, three-speed
49WL - 45ci (750cc) flathead V-twin, three-speed
49E Hydra-Glide - 61ci (**1000cc**) OHV V-twin, medium-compression, four-speed
49EL Hydra-Glide Sport - 61ci OHV V-twin, high-compression, four-speed
49ES - 61ci OHV V-twin, medium-compression, four-speed, sidecar gearing
49EP - 61ci OHV V-twin, medium-compression, four-speed, spring fork, sidecar gearing
49ELP Sport - 61ci OHV V-twin, medium-compression, four-speed, spring fork
49F Hydra-Glide - 74ci (**1200cc**) OHV V-twin, medium-compression, four-speed
49FL Hydra-Glide - 74ci OHV V-twin, high-compression, four-speed

OPPOSITE: The 1948 Servicar was little different to the 1938 version, with the same 45-ci (750-cc) side-valve twin and 3-speed hand-change gearbox.

RIGHT: The first Panhead of 1948.

BELOW: The Hydra-Glide of 1949/50.

49FLP Sport - 74ci OHV V-twin, high-compression, four-speed
49FS - 74ci OHV V-twin, medium-compression, four-speed, sidecar gearing
49FP - 74ci OHV V-twin, medium-compression, four-speed, spring fork sidecar gearing
49G Servicar - 45ci flathead V-win, three-speed+reverse

Prices
S - $325
WL - $590
E - $735
EL - $735
ES - $735
F - $750
FL - $750
FS - $750
Servicar - $860

Production
S - 7,291
WL - 2,289
WLA - 436
EL - 3,419
ELP - 99
ES - 177
FL - 8,014
FLP - 486
FS - 490
Servicars - 1,039
TOTAL - 23,740

Colours
Single - Black
Twins - Brilliant Black, Peacock Bue, Burgundy, Police Blue, Metallic Congo Green

Specifications
S 7ci Single - No major changes
W 45ci V-twin - No major changes
E/F 61/74ci Twins - Hydra-Glide hydraulically-damped forks (includes larger front brake and headlamp, redesigned bars), spring forks still optional
Servicars - No major changes

The 1952 WR racer, the only one in existence. Thanks to the tuning genius of Tom Sifton and others, it was a competitive machine for several years.

1950

The Range

50S - 7ci (125cc) two-stroke single, three-speed
50WL - 45ci (750cc) flathead V-twin, three-speed
50E Hydra-Glide - 61ci (**1000cc**) OHV V-twin, medium-compression, four-speed
50EL Hydra-Glide Sport - 61ci OHV V-twin, high-compression, four-speed
50ES - 61ci OHV V-twin, medium-compression, four-speed, adjustable-rake forks, sidecar gearing
50F Hydra-Glide - 74ci (**1200cc**) OHV V-twin, medium-compression, four-speed
50FL Hydra-Glide - 74ci OHV V-twin, high-compression, four-speed
50FS Hydra-Glide - 74ci OHV V-twin, medium-compression, four-speed, adjustable-rake forks, sidecar gearing
50G Servicar - 45ci flathead V-win, three-speed+reverse

Prices

S - $325
WL - $590
E - $735
EL - $735
ES - $735
F - $750
FL - $750
FS - $750
Servicar - $860

Production

S - 4,708
WL - 1,108
WLA - 15
EL - 2,046
ES - 177
FL - 7,407
FS - 544
Servicars - 1,003
TOTAL - 17,008

Colours

Single - Brilliant Black, Flight Red, Sportsman Yellow, Riviera Blue
Twins - Brilliant Black, Riviera Blue, Ruby Red, Police Silver (police only), White, Flight Red, Metallic Green, Azure Blue

Specifications

S 7ci Single - No major changes
W 45ci V-twin - No major changes
E/F 61/74ci Twins - Redesigned cylinder heads (10% power increase)
Servicars - No major changes

1951

The Range

51S - 7ci (125cc) two-stroke single, three-speed
51WL - 45ci (750cc) flathead V-twin, three-speed
51EL Hydra-Glide Sport - 61ci (**1000cc**) OHV V-twin, high-compression, four-speed
51ELS - 61ci OHV V-twin, medium-compression, four-speed, sidecar gearing
51FL Hydra-Glide Sport - 74ci (**1200cc**) OHV V-twin, high-compression, four-speed
51FLS Hydra-Glide - 74ci OHV V-twin, medium-compression, four-speed, sidecar model
51G Servicar - 45ci flathead V-win, three-speed+reverse

Prices

S - $365
WL - $730
EL - $885
ELS - $885
F - $750
FL - $900
FLS - $900
Servicar - $1,095

Production

S - 5,101

WL - 1,044
WLA - 1
EL - 1,532
ES - 76
FL - 6,560
FS - 135
Servicars - 778
TOTAL - 15,227

Colours

Single - Persian Red, Sportsman Yellow, Rio Blue, Metallic Blue, Metallic Green
Twins - Brilliant Black, Rio Blue, Persian Red, Police Silver (police only), White, Metallic Blue, Metallic Green

Specifications

S 7ci Single - Roller-bearing small-end bearing, new speedometer, other minor changes
W 45ci V-twin - Fixed jet Linkert carburettor
E/F 61/74ci Twins - Several minor changes
Servicars - Hydraulic rear brake, 16in (406mm) rear wheels

1952

The Range

52S - 7ci (125cc) two-stroke single, three-speed
52K Sport - 45ci (750cc) flathead V-twin, four-speed
52EL Hydra-Glide Sport - 61ci (**1000cc**) OHV V-twin, four-speed
52ELS - 61ci OHV V-twin, four-speed, sidecar model
52ELF - 61ci OHV V-twin, four-speed, foot-change
52FL Hydra-Glide Sport - 74ci (**1200cc**) OHV V-twin, four-speed
52FLS Hydra-Glide - 74ci OHV V-twin, four-speed, sidecar model
52FLF Hydra-Glide - 74ci OHV V-twin, four-speed, foot-change
52G Servicar - 45ci flathead V-win, three-speed+reverse

Prices

WL - $730
K - $865
EL - $955
ELS - $955
ELF - $955
F - $750
FL - $970
FLS - $970
FLF - $970
Servicar - $1175

Production

S - 4,576
K - 1,970
EL - 918
ES - 42
FL - 5,554
FLS - 186
Servicars - 1,047
TOTAL - 14,293

Colours

Single - Tropical Green, Rio Blue, Persian Red, Marine Blue Metallic
K - Black, Rio Blue, Persian Red, Bronco Bronze
E/F Twins - Brilliant Black, Rio Blue, Tropical Green, Persian Red, Police Silver (police only), Marine Blue Metallic, Bronco Bronze Metallic, White

Specifications

S 7ci Single - Many minor changes
K 45ci V-twin - New 45ci flathead V-twin, replacing W-series, 2.75in x 3.813in (70 x 97mm) bore x stroke, unit-construction, aluminium-alloy cylinder heads, alloy pistons, 30hp, four-speed gearbox, foot-change, hand-clutch, swinging-arm rear suspension, hydraulic front forks, six-volt electrics, 19in (483mm) wheels
E/F 61/74ci Twins - Foot-change optional, many minor changes
Servicars - No major changes

1953

The Range

53ST - 10ci (165cc) two-stroke single, three-speed
53K Sport - 45ci (**750cc**) flathead V-twin, four-speed
53KK - 45ci flathead V-twin, four-speed
53FL Hydra-Glide Sport - 74ci (**1200cc**) OHV V-twin, four-speed
53FLF Hydra-Glide - 74ci OHV V-twin,, four-speed, foot-change
53FLE Hydra-Glide - 74ci OHV V-twin, four-speed, sidecar model
53FLEF Hydra-Glide - 74ci OHV V-twin, four-speed, foot-change, sidecar model
53G Servicar - 45ci flathead V-win, three-speed+reverse

Prices

ST - $405
K - $875
FL - $1,000
FLF - $1,000
FLE - $1,000
FLEF - $1,000
Servicar - $1,190

Production

ST - 4,225
K - 1,723
FL - 1,986
FLF - 3,351
Servicars - 1,146
TOTAL - 12,431

Colours

Single - Pepper Red, Glacier Blue, Forest Green, Glamour Green
Twins - Pepper Red, Forest Green, Silver (police only), Glacier Blue, Brilliant Black, White, Cavalier Brown, Glamour Green, Cadillac Grey/Azure Blue (Motor Maids special)

Specifications

ST 10ci Single - Replaces S-series, 165cc two-stroke, .375in x 2.281in (9.5 x 58mm) bore x stroke, Alemite grease fittings, revised gearing, 5hp restrictor kit available for those with restricted licence
Many minor changes
K 45ci V-twin - KK has KR racing spec cams, ports and ball-bearing crankshaft
F 74ci Twins - Redesigned hydraulic tappets for more consistent valve timing, FLE 74ci motor replaces 61ci, with milder cam and smaller carburettor
Servicars - No major changes

1954

The Range

54ST - 10ci (**165cc**) two-stroke single, three-speed
54STU - 10ci two-stroke single, three-speed
54KH Sport - 55ci (900cc) flathead V-twin, four-speed
54FL Hydra-Glide Sport - 74ci (**1200cc**) OHV V-twin, four-speed
54FLF Hydra-Glide - 74ci OHV V-twin, four-speed, foot-change
54FLE Hydra-Glide - 74ci OHV V-twin, four-speed, sidecar model
54FLEF Hydra-Glide - 74ci OHV V-twin, four-speed, foot-change, sidecar model
54G Servicar - 45ci flathead V-win, three-speed+reverse

Prices

ST - $405
STU - $405
KH - $925
FL - $1,015
FLF - $1,015
FLE - $1,015
FLEF - $1,015
Servicar - $1,240

Production

ST - 2,835
KH - 1,579
FL - 4,757
Servicars - 1,397
TOTAL - 10,568

Colours

Pepper Red, Daytona Ivory, Glacier Blue, Anniversary Yellow, Forest Green, Black, White (police only), Silver (police only), Cadillac Grey/Azure Blue (Motor Maids special)

The little 7-ci (125-cc) Hummer was based on a German DKW design. This 1954 bike was built in Harley-Davidson's 50th Anniversary year.

Specifications
ST 10ci Single - STU is 5hp restricted version
KH 55ci V-twin - Replaces 45ci K, 55ci flathead V-twin, 2.745in x 4.562in (70 x 116mm) bore x stroke, 6.8:1 compression ratio, 38hp (12% power increase), aluminium pistons, gearbox and clutch redesigned to suit, revised geometry
F 74ci Twins - No major changes
Servicars - No major changes

1955

The Range
55B Hummer - 7ci (125cc) two-stroke single, three-speed
55ST- 10ci (**165cc**) two-stroke single, three-speed
55STU - 10ci two-stroke single, three-speed
55KH Sport - 55ci (**900cc**) flathead V-twin, four-speed
55KHK Super Sport - 55ci
55FL Hydra-Glide Sport - 74ci (**1200cc**) OHV V-twin, four-speed
55FLF Hydra-Glide Sport - 74ci OHV V-twin, four-speed, foot-change
55FLE Hydra-Glide - 74ci OHV V-twin, four-speed, sidecar model
55FLEF Hydra-Glide - 74ci OHV V-twin, four-speed, foot-change, sidecar model
55FLH Hydra-Glide Super Sport - 74ci OHV V-twin, four-speed
55FLHF Hydra-Glide Super Sport - 74ci OHV V-twin, four-speed, foot-change
55G Servicar - 45ci (750cc) flathead V-win, three-speed+reverse

Prices
ST - $405
STU - $405
KH - $925
KHK - $995
FL - $1,015
FLF - $1,015
FLE - $1,015
FLEF - $1,015
FLH - $1,083
FLHF - $1,083
Servicar - $1,240

Production
ST - 2,263
B - 1,040
KH - 1,065
KR Racers -135
FL - 5,142
Servicars - 1,041
TOTAL - 10,686

Colours
Black, Pepper Red, Aztec Brown, Anniversary Yellow, Atomic Blue, Silver (police only), White (police only), Hollywood Green

Specifications
B 7ci Single - New model, 125cc two-stroke single, $2^{1}/_{16}$ x $2^{9}/_{32}$in (52 x 60mm) bore x stroke, 6.6:1 compression, 1.7hp, three-speed gearbox, multi-plate clutch
ST/STU 10ci Single - No major changes

KH 55ci V-twin - KHK Super Sport with sports cams and polished cylinder heads
F 74ci Twins - FLH Super Sport, new model, 8:1 compression ratio, polished and flowed intake ports, many minor changes
Servicars - No major changes

1956

The Range
56B Hummer - 7ci (125cc) two-stroke single, three-speed
56ST- 10ci (**165cc**) two-stroke single, three-speed
56STU - 10ci two-stroke single, restricted, three-speed
56KH Sport - 55ci (**900cc**) flathead V-twin, four-speed
56KHK Super Sport - 55ci
56FL Hydra-Glide Sport - 74ci (**1200cc**) OHV V-twin, four-speed
56FLF Hydra-Glide Sport - 74ci OHV V-twin, four-speed, foot-change
56FLE Hydra-Glide - 74ci OHV V-twin, four-speed, sidecar model
56FLEF Hydra-Glide - 74ci OHV V-twin, four-speed, foot-change, sidecar model
56FLH Hydra-Glide Super Sport - 74ci OHV V-twin, high-compression, four-speed
56FLHF Hydra-Glide Super Sport - 74ci OHV V-twin, high-compression, four-speed, foot-change
56G Servicar - 45ci (750cc) flathead V-win, three-speed+reverse

Prices
B - $320
ST - $405
STU - $405
KH - $935
KHK - $1,003
FL - $1,055
FLF - $1,055
FLE - $1,055
FLEF - $1,055
FLH - $1,123
FLHF - $1,123
Servicar - $1,240

Production
ST - 2,219
B - 1,384
KH - 539
KHK - 714
KR Racers - 60
FL/FLH - 5,806
Servicars - 1,203
TOTAL - 11,925

Colours
B Single - Pepper Red, Atomic Blue
ST Singles - Pepper Red, Atomic Blue, Champion Yellow, Tangerine
Twins - Pepper Red, Black, Atomic Blue, Champion Yellow, Silver (police only), White (police only), Flamboyant Metallic Green

Specifications
B 7ci Single - No major changes
ST/STU 10ci Single - No major changes
KH 55ci V-twin - Lower saddle, redesigned oil pump, redesigned gearbox
F 74ci Twins - Higher lift cams for FLH, many minor changes
Servicars - Some minor changes

1955, and the Big Twin has yet to acquire rear suspension or a genuine hardtail. But Hydra-Glide forks on big soft tyres and a sprung saddle were enough for some.

ABOVE LEFT: The 1956 KH was substantially a new bike, with modern chassis and four-speed foot-change gearbox, but a side-valve motor.
ABOVE : Duo Glide brought rear suspension at last.

1957

The Range
57B Hummer - 7ci (125cc) two-stroke single, three-speed
57ST - 10ci (**165cc**) two-stroke single, three-speed
57STU - 10ci two-stroke single, three-speed
57XL Sportster - 54ci (883cc) OHV V-twin, four-speed
57FL Hydra-Glide Sport - 74ci (**1200cc**) OHV V-twin, four-speed
57FLF Hydra-Glide Sport - 74ci OHV V-twin, four-speed, foot-change
57FLH Hydra-Glide Super Sport - 74ci OHV V-twin, high-compression, four-speed
57FLHF Hydra-Glide Super Sport - 74ci OHV V-twin, high-compression, four-speed, foot-change
57G Servicar - 45ci (750cc) flathead V-win, three-speed+reverse

Prices
B - $356
ST - $445
STU - $445
XL - $1,103
FL - $1,167
FLF - $1,167
FLH - $1,243
FLHF - $1,243
Servicar - $1,367

Production
ST - 2,401
B - 1,350
KH - 90
KR Racers - 29
XL - 1,983
FL/FLH - 5,616
Servicars - 1,192
TOTAL - 12,661

Colours
Singles - Pepper Red, High Fire, Skyline Blue, Birch White
Twins - Pepper Red, Black, Skyline Blue, Birch White, Metallic Midnight Blue
Servicars - Pepper Red, Black, Skyline Blue, Silver (police only), Metallic Midnight Blue

Specifications
Hummer 7ci Single - No major changes
ST/STU 10ci Single - No major changes
XL 54ci Twins - New model, 54ci V-twin, 3in x 3.812in (76 x 97mm) bore x stroke, 7.5:1 compression ratio, unit construction, dry sump, cast-iron cylinders and heads, four-speed gearbox, foot-change, 18in (457mm) wheels, telescopic front forks, rear swinging arm
F 74ci Twins - Some minor changes
Servicars - Some minor changes

1958

The Range
58B Hummer - 7ci (125cc) two-stroke single, three-speed
58ST - 10ci (**165cc**) two-stroke single, three-speed
58STU - 10ci two-stroke single, restricted, three-speed
58XL Sportster - 54ci (**883cc**) OHV V-twin, medium-compression, four-speed
58XLH Sportster - 54ci OHV V-twin, high-compression, four-speed
58XLC Sportster Racing - 54ci OHV V-twin, high-compression, four-speed
58XLCH Sportster Super Sport - 54ci OHV V-twin, high-compression, four-speed
58FL Duo Glide Sport - 74ci (**1200cc**) OHV V-twin, four-speed
58FLF Duo Glide Sport - 74ci OHV V-twin, four-speed, foot-change
58FLH Duo Glide Super Sport - 74ci

OHV V-twin, high-compression, four-speed
58FLHF Duo Glide Super Sport - 74ci OHV V-twin, high-compression, four-speed, foot-change
58G Servicar - 45ci (750cc) flathead V-win, three-speed+reverse

Prices
B - $375
ST - $465
STU - $465
XL - $1,155
FL - $1,255
FLF - $1,255
FLH - $1,320
FLHF - $1,320
Servicar - $1,450

Production
ST - 2,445
B - 1,677
KR Racers - 9
XL - 579
XLH - 711
XLCH - 239
FL/FLH - 6,038
Servicars - 1,192
TOTAL - 12,890

Colours
Singles - Skyline Blue, Tropical Coral, Calypso Red, Black
XL Sportster - Skyline Blue, Sabre Grey Metallic, Calypso Red, Black
Twins - Calypso Red, Black, Skyline Blue, Sabre Grey Metallic, Police Silver, Birch White (police only)
Servicars - Calypso Red, Black, Skyline Blue, Birch White, Police Silver

Specifications
B 7ci Single - No major changes
ST/STU 10ci Single - No major changes
XL 54ci Twins - XLC/XLCH new models, off-road versions of XL/SLH, with 2.25gal (1 litre) 'peanut' tank, magneto ignition, no silencers, no lights. Otherwise, no major changes
F 74ci Twins - No major changes
Servicars - Hydra-Glide front forks standard

1959

The Range
59B Hummer - 7ci (125cc) two-stroke single, three-speed
59ST - 10ci (**165cc**) two-stroke single, three-speed
59STU - 10ci two-stroke single, restricted, three-speed
59XL Sportster - 54ci (**883cc**) OHV V-twin, medium-compression, four-speed
5XLH Sportster - 54ci OHV V-twin, high-compression, four-speed
59XLC Sportster Racing - 54ci OHV V-twin, high-compression, four-speed
59XLCH Sportster Super Sport - 54ci OHV V-twin, high-compression, four-speed
59FL Duo Glide Sport - 74ci (**1200cc**) OHV V-twin, four-speed
59FLF Duo Glide Sport - 74ci OHV V-twin, four-speed, foot-change
59FLH Duo Glide Super Sport - 74ci OHV V-twin, high-compression, four-speed
59FLHF Duo Glide Super Sport - 74ci OHV V-twin, high-compression, four-speed, foot-change
59G Servicar - 45ci (750cc) flathead V-win, three-speed+reverse

Prices
B - $385
ST - $475
STU - $475
XL - $1,175
XLH - $1,200
XLCH - $1,285
FL - $1,280
FLF - $1,280
FLH - $1,345
FLHF - $1,345
Servicar - $1,500

Production
ST - 2,311
B - 1,285
KR Racers - 33
XL - 42
XLH - 947
XLCH - 1,059
FL/FLH - 5,767
Servicars - 812
TOTAL - 12,256

Colours
Singles - Skyline Blue, Calypso Red, Hi-Fi Turquoise, Hi-Fi Red
XL Sportster - Skyline Blue, Calypso Red, Black, Hi-Fi Turquoise, Hi-Fi Red
Twins - Calypso Red, Black, Skyline Blue, Hi-Fi Turquoise, Hi-Fi Red, Police Silver, Birch White (police only)
Servicars - Calypso Red, Black, Skyline Blue, Birch White, Police Silver

Specifications
B 125cc Single - No major changes
ST/STU 165cc Single - No major changes
XL Sportsters - XLH has high-lift intake cams, XLCH now road legal, with lights, battery/generator electrics, 2-into-1 exhaust with silencer
FL Duo Glides - Duo Glide new model, swinging-arm rear suspension with twin hydraulic shock absorbers, redesigned frame, hydraulic rear brake, many minor changes
Servicars - No major changes

The Sportster: This is a 1959 XLCH with the small tank (from the Hummer) and high-level pipes and was the start of a new and highly successful family of Harley-Davidsons.

1960

The Range
60A Topper - 10ci (**165cc**) two-stroke single, automatic transmission
60AU Topper - 10ci two-stroke single, 5hp restricted, automatic
60BT Super 10 - 10ci two-stroke single, three-speed
60BTU Super 10 - 10ci two-stroke single, 5hp restricted, three-speed
60XLH Sportster - 54ci (**883cc**) OHV V-twin, four-speed
60XLCH Sportster - 54ci OHV V-twin, four-speed
60FL Duo Glide Sport - 74ci (**1200cc**) OHV V-twin, four-speed
60FLF Duo Glide Sport - 74ci OHV V-twin, four-speed, foot-change
60FLH Duo-Glide Super Sport - 74ci OHV V-twin, high-compression, four-speed
60FLHF Duo Glide Super Sport - 74ci OHV V-twin, high-compression, four-speed, foot-change
60G Servicar - 45ci (750cc) flathead V-win, three-speed+reverse

Prices
Topper AH - $430
TopperAU - $430
Super 10 BT - $455
Super 10 BTU - $465
Sportster XLH - $1,225
Sportster - $1,310
Duo Glide FL - $1,310
Duo Glide FLF - $1,310
Duo Glide FLH - $1,375
Duo Glide FLHF - $1,375
Servicar - $1,530

Production
Topper - 3,801
Super 10 - 2,488
Sportster - 2,765
Duo Glide - 5,967
Servicars - 707
TOTAL - 15,728

Colours
Topper - Pepper Red, Strato Blue, Granada Green
Super 10 - Pepper Red, Hi-Fi Red, Hi-Fi Green, Hi-Fi Blue
Sportster - Skyline Blue, Black, Hi-Fi Red, Hi-Fi Green, Hi-Fi Blue
Twins - Black, Skyline Blue, Hi-Fi Red, Hi-Fi Green, Hi-Fi Blue, Police Silver, Birch White (police only)
Servicars - Black, Skyline Blue, Birch White, Police Silver

Specifications
A/AU Topper scooter - New model, 165cc two-stroke single, 2.375in x 2.281in (60 x 58mm) bore x stroke, 6.6:1 compression, reed intake valve, aluminium cylinder head, rubber mountings, automatic belt drive and chain final drive, leading-link front suspension, swinging-arm rear suspension, recoil starter, AU is 5hp restricted version
Super 10 - New model update of B single, now 10ci two-stroke single, as Topper, 6hp (BT) and 5hp (BTU)
XL Sportster - No major changes
FL Duo Glides - New headlamp nacelle, some minor changes
Servicars - New headlamp nacelle

ABOVE: Not many people know that Harley-Davidson made a scooter. The 10-ci (165-cc) Topper, this one from 1960, was a belated attempt to cash in on the growing scooter market. However it had none of the style of a Vespa or Lambretta, but the Topper is a very rare beast today.

OPPOSITE: A 1960 Duo Glide, set up for touring with screen and panniers or saddlebags.

1961

The Range

61AH Topper - 10ci (**165cc**) two-stroke single, automatic transmission
61AU Topper - 10ci two-stroke single, 5hp restricted, automatic
61BT Super 10 - 10ci two-stroke single, three-speed
61BTU Super 10 - 10ci two-stroke single, 5hp restricted, three-speed
61C Sprint - 15ci (246cc) OHV single, four-speed
61XLH Sportster - 54ci (**883cc**) OHV V-twin, four-speed
61XLCH Sportster - 54ci OHV V-twin, four-speed
61FL Duo Glide Sport - 74ci (**1200cc**) OHV V-twin, four-speed
61FLF Duo Glide Sport - 74ci OHV V-twin, four-speed, foot-change
61FLH Duo Glide Super Sport - 74ci OHV V-twin, high-compression, four-speed
61FLHF Duo Glide Super Sport - 74ci OHV V-twin, high-compression, four-speed, foot-change
61G Servicar - 45ci (750cc) flathead V-win, three-speed+reverse

Prices

Topper AU - $445
Super 10 BT - $465
Super 10 BTU - $465
Sportster XLH - $1,250
Sportster XLCH - $1,335
Duo Glide FL - $1,335
Duo Glide FLF - $1,335
Duo Glide FLH - $1,400
Duo Glide FLHF - $1,400
Servicar - $1,530

Production

Topper - 1,341
Super 10 - 1,527
Sportster - 2,014
Duo Glide - 4,927
Servicars - 628
TOTAL - 10,437

Colours

Topper - Strato Blue
Super 10 - Skyline Blue, Hi-Fi Red, Hi-Fi Green, Hi-Fi Blue, Birch White
Sportster - Pepper Red, Black, Hi-Fi Red, Hi-Fi Green, Hi-Fi Blue
Duo Glide - Black, Pepper Red, Hi-Fi Red, Hi-Fi Green, Hi-Fi Blue, Police Silver, Birch White (police only)
Servicars - Black, Skyline Blue, Birch White, Pepper Red, Police Silver

Specifications

AH/AU Topper - AH has high-compression 9hp motor, sidecar and utility box optional
Super 10 - 16in (406mm) wheels
Sprint - New model, Aermacchi Sprint made in Italy, rebadged as Harley-Davidson. 15ci OHV four-stroke single, 2.6 x 2.83in (66 x 72mm), 18bhp/7,500rpm, four-speed, motor hung from single strut steel frame
XL Sportster - No major changes
FL Duo Glides - 'Single-fire' ignition (separate points and coil for each cylinder)
Servicars - No major changes

ABOVE LEFT: The Super 10 was a 10-ci (165-cc) development of the Hummer, and still sold reasonably well despite the age of the basic design.

ABOVE: A 1961 Duo Glide festooned with period extras, just the sort of thing the average long distance rider would fit: glassfibre panniers (starting to replace traditional leather), screen, pillion grab handle, whitewall tyres and screen. In other words, a dresser.

1962

The Range

62AH Topper - 10ci (**165cc**) two-stroke single, automatic transmission
62AU Topper - 10ci two-stroke single, 5hp restricted, automatic
62BT Pacer - 11ci (**175cc**) two-stroke single, three-speed
62BTF Ranger - 10ci two-stroke single, three-speed
62BTH Scat - 11ci two-stroke single, three-speed
62BTU Pacer - 10ci two-stroke single, 5hp restricted, three-speed
62C Sprint - 15ci (**246cc**) OHV single, four-speed
62H Sprint H - 15ci OHV single, high compression, four-speed
62XLH Sportster - 54ci (**883cc**) OHV V-twin, four-speed
62XLCH Sportster - 54ci OHV V-twin, four-speed
62FL Duo Glide Sport - 74ci (**1200cc**) OHV V-twin, four-speed
62FLF Duo Glide Sport - 74ci OHV V-twin, four-speed, foot-change
62FLH Duo Glide Super Sport - 74ci OHV V-twin, high-compression, four-speed
62FLHF Duo Glide Super Sport - 74ci OHV V-twin, high-compression, four-speed, foot-change
62G Servicar - 45ci (750cc) flathead V-win, three-speed+reverse

Prices

Topper AH - $445
Topper AU - $445
Pacer BT - $465
Ranger BTF - $440
Scat BTH - $475
Pacer BTU - $465
Sportster XLH - $1,250
Sportster XLCH - $1,335
Duo Glide FL - $1,335
Duo Glide FLF - $1,335
Duo Glide FLH - $1,400
Duo Glide FLHF - $1,400
Servicar - $1,555

Production

Pacer/Ranger/Scat - 1,983
Sportster - 1,998
Duo-Glide - 5,184
Servicars - 703
TOTAL - 9,868

Colours

Topper - Birch White, Tango Red, Granada Green
Pacer/Ranger/Scat - Tango Red, Skyline Blue, Hi-Fi Red, Hi-Fi Purple
Sportster - Black, Tango Red, Hi-Fi Red, Hi-Fi Purple, Hi-Fi Blue
Duo Glide - Black, Tango Red, Hi-Fi Red, Hi-Fi Purple, Hi-Fi Blue
Servicars - Black, Birch White, Tango Red, Skyline Blue, Police Silver (police only)

Specifications

AH/AU Topper - No major changes
BT Pacer - New update of Super 10, 11ci two-stroke single, longer stroke, higher compression (7.65:1), BTH Scat is on/off-road version, BTF Ranger off-road, BTU Pacer 5hp restricted - BTU and BTF continue with 10ci single
Sprint H - New model based on Sprint. 9.2:1 compression, 19.5bhp/6,700rpm, off-road styling with short mudguards, high-level exhaust, semi-knobbly tyres
XL Sportster - No major changes
FL Duo Glides - No major changes
Servicars - No major changes

1963

The Range

63AH Topper - 10ci (**165cc**) two-stroke single, automatic transmission
63AU Topper - 10ci two-stroke single, 5hp restricted, automatic
63BT Pacer - 11ci (**175cc**) two-stroke single, three-speed
63BTH Scat - 11ci two-stroke single, three-speed
63BTU Pacer - 10ci two-stroke single, 5hp restricted, three-speed
63C Sprint - 15ci (**246cc**) OHV single, four-speed
63H Sprint H - 15ci OHV single, high-compression, four-speed
63XLH Sportster - 54ci (**883cc**) OHV V-twin, four-speed
63XLCH Sportster - 54ci OHV V-twin, four-speed
63FL Duo Glide Sport - 74ci (**1200cc**) OHV V-twin, four-speed
63FLF Duo Glide Sport - 74ci OHV V-twin, four-speed, foot-change
63FLH Duo Glide Super Sport - 74ci OHV V-twin, high-compression, four-speed
63FLHF Duo Glide Super Sport - 74ci OHV V-twin, high-compression, four-speed, foot-change
63G Servicar - 45ci (750cc) flathead V-twin, three-speed+reverse

Prices

Topper AH - $460
Topper AU - $460
Pacer BT - $485
Scat BTH - $495
Pacer BTU - $485
Sportster XLH - $1,270
Sportster XLCH - $1,355
Duo-Glide FL - $1,360
Duo-Glide FLF - $1,360
Duo-Glide FLH - $1,425
Duo-Glide FLHF - $1,425
Servicar - $1,590

Production

Topper - 978
Pacer/Ranger/Scat - 1,740
Sportster - 1,433
Duo Glide - 4,246
TOTAL - 8,397

Colours

Topper - Black, Tango Red, White
Pacer/Scat - Tango Red, Black, Hi-Fi Red, Hi-Fi Turquoise
Sportster - Black, Tango Red, Horizon Metallic Blue, Hi-Fi Turquoise, Hi-Fi Red, Hi-Fi Purple
Duo Glide - Black, Tango Red, Horizon Metallic Blue, Hi-Fi Red, Hi-Fi Purple, Hi-Fi Turquoise
Servicars - Black, Birch White, Tango Red, Horizon Metallic Blue, Police Silver (police only)

Specifications

AH/AU Topper - No major changes
BT/BTH/BTU - Swinging-arm rear suspension, BTU has restricterd 11ci single
Sprint H - Power increase to 25bhp/8,700rpm, 1.06in (27mm) Dell'Orto carburettor, megaphone exhaust
XL Sportster - No major changes
FL Duo-Glides - No major changes
Servicars - No major changes

1964

The Range

64AH Topper - 10ci (**165cc**) two-stroke single, automatic transmission
64AU Topper - 10ci two-stroke single, 5hp restricted, automatic
64BT Pacer - 11ci (**175cc**) two-stroke single, three-speed
64BTH Scat - 11ci two-stroke single, three-speed
64BTU Pacer - 10ci two-stroke single, 5hp restricted, three-speed
64C Sprint - 15ci (**246cc**) OHV single, four-speed
64H Sprint H - 15ci OHV single, high-compression, four-speed
64XLH Sportster - 54ci (**883cc**) OHV V-twin, four-speed
64XLCH Sportster - 54ci OHV V-twin, four-speed
64FL Duo Glide Sport - 74ci (**1200cc**) OHV V-twin, four-speed
64FLF Duo Glide Sport - 74ci OHV V-

twin, four-speed, foot-change
64FLH Duo Glide Super Sport - 74ci OHV V-twin, high-compression, four-speed
64FLHF Duo Glide Super Sport - 74ci OHV V-twin, high-compression, four-speed, foot-change
64GE Servicar - 45ci (750cc) flathead V-win, three-speed+reverse

Prices
Topper AH - $470
Topper AU - $470
Pacer BT - $495
Scat BTH - $505
Pacer BTU - $495
Sportster XLH - $1,295
Sportster XLCH - $1,360
Duo-Glide FL - $1,385
Duo-Glide FLF - $1,385
Duo-Glide FLH - $1,450
Duo-Glide FLHF - $1,450
Servicar - $1,628

Production
Topper - 825
Pacer/Scat - 2,760
Sportster - 1,433
Duo Glide - 5,500
Servicar - 725
TOTAL - 11,243

Colours
Topper - Black, Fiesta Red
Pacer/Scat - Black, Fiesta Red
Sportster - Black, Fiesta Red, Hi-Fi Red, Hi-Fi Blue
Duo Glide - Black, Fiesta Red, Hi-Fi Red, Hi-Fi Blue
Servicars - Black, White, Fiesta Red, Police Silver (police only)

Specifications
AH/AU Topper - No major changes
BT/BTH/BTU - No major changes
XL Sportster - Full-width aluminium drum front brake
FL Duo Glides - No major changes
Servicars - Electric starter, 12-volt electrics

1965

The Range
65M 50 - 3ci (49cc) two-stroke single, three-speed
65AH Topper - 10ci (165cc) two-stroke single, automatic transmission
65BT Pacer - 11ci (**175cc**) two-stroke single, three-speed
65BTH Scat - 11ci two-stroke single, three-speed
65C Sprint - 15ci (**246cc**) OHV single, four-speed
65H Sprint H - 15ci OHV single, high-compression, four-speed
65XLH Sportster - 54ci (**883cc**) OHV V-twin, four-speed
65XLCH Sportster - 54ci OHV V-twin, four-speed
65FLB Electra Glide Sport - 74ci (**1200cc**) OHV V-twin, four-speed
65FLFB Electra Glide Sport - 74ci OHV V-twin, four-speed, foot-change
65FLHB Electra Glide Super Sport - 74ci OHV V-twin, high-compression, four-speed
65FLHFB Electra Glide Super Sport - 74ci OHV V-twin, high-compression, four-speed, foot-change
65GE Servicar - 45ci (750cc) flathead V-win, three-speed+reverse

Production
Topper - 500
Pacer/Scat - 1,250
Sportster - 3,770
Electra Glide - 6,930
Servicar - 625
TOTAL - 13,075

Colours
M-50 - Holiday Red
Topper - Pacific Blue, Holiday Red
Pacer/Scat - Black, Holiday Red, Pacific Blue
Sportster - Black, Holiday Red, Hi-Fi Red, Hi-Fi Blue
Electra Glide - Black, Holiday Red, Birch White (police only), Police Silver (police only), Hi-Fi Red, Hi-Fi Blue
Servicars - Black, Birch White, Holiday Red, Police Silver (police only)

Specifications
M-50 - New model, Aermacchi-built two-stroke moped, three-speed, motorcycle styling
AH Topper - No major changes
BT/BTH - No major changes
XL Sportster - 12-volt electrics, auto-advance ignition
FL Electra Glides - Electric start, 12-volt electrics, 5-gal (22.7-litre) tank, auto-advance ignition
Servicars - No major changes

1966

The Range
66M 50 - 3ci (**49cc**) two-stroke single, three-speed
66MS 50 Sport - 3ci two-stroke single, three-speed
66BTH Bobcat - 11ci (175cc) two-stroke single, three-speed
66C Sprint - 15ci (**246cc**) OHV single, four-speed
66H Sprint H - 15ci OHV single, high-compression, four-speed
66XLH Sportster - 54ci (**883cc**) OHV V-twin, four-speed
66XLCH Sportster - 54ci OHV V-twin, four-speed
66FLB Electra Glide Sport - 74ci (**1200cc**) OHV V-twin, four-speed
66FLFB Electra Glide Sport - 74ci OHV V-twin, four-speed, foot-change
66FLHB Electra Glide Super Sport - 74ci OHV V-twin, high-compression, four-speed

66FLHFB Electra Glide Super Sport - 74ci OHV V-twin, high-compression, four-speed, foot-change

Prices

Bobcat - $515
Sportster XLH - $1,411
Sportster XLCH - $1,415
Electra-Glide FLB - $1,545
Electra-Glide FLFB - $1,545
Electra-Glide FLHB - $1,610
Electra-Glide FLHFB - $1,610

Production

Bobcat - 1,150
Sportster -4,800
Electra Glide - 7,800
Servicar - 625
TOTAL - 14,375

Colours

Bobcat - Holiday Red, Indigo Metallic, Sparkling Burgundy
Electra Glide - Black, Indigo Metallic, Sparkling Burgundy, Birch White (police only), Hi-Fi Blue
Servicars - Black, Birch White, Holiday Red

Specifications

MS50 Sport - Restyled version of M50, racing-style fuel tank, flat seat
BTH Bobcat - Restyle of BT Pacer, mechanically unchanged
XL Sportster - 15% power increase, Tillotson carburettor
FL Electra Glides - New 'Shovelhead' V-twin, new cylinders, aluminium cylinder heads, Sportster-style rocker boxes, Linkert, then Tillotson carburettor
Servicar - No major changes

1967

The Range

67M 50 - 3ci (49cc) two-stroke single, three-speed
67M 65 - 5ci (**65cc**) two-stroke single, three-speed

ABOVE: Electra Glide: Harley's best known name was born in 1965 with the unveiling of an electric- start Duo Glide. Apart from that and a 12-volt system to cope with the rigours of turning over 74- cubic inches of compression on a winter's morning, it was no different. However, the extra electrics added a massive 75lb (34kg) to the weight of the machine. The Panhead motor (seen here) was only offered in the Electra Glide for 1965.

OPPOSITE: 1964, and the Duo Glide is still the standard American touring motorcycle, still in kickstart form only.

67MS 65 - 5ci two-stroke single, three-speed
67H Sprint H - 15ci (**246cc**) OHV single, four-speed
67SS Sprint SS - 15ci OHV single, four-speed
67XLH Sportster - 54ci (**883cc**) OHV V-twin, four-speed
67XLCH Sportster - 54ci OHV V-twin, four-speed
67FLB Electra Glide Sport - 74ci (**1200cc**) OHV V-twin, four-speed
67FLFB Electra Glide Sport - 74ci OHV V-twin, four-speed, foot-change
67FLHB Electra Glide Super Sport - 74ci OHV V-twin, high-compression, four-speed
67FLHFB Electra Glide Super Sport - 74ci OHV V-twin, high-compression, four-speed, foot-change
67GE Servicar - 45ci (750cc) flathead V-twin, three-speed+reverse

Prices

Sportster XLH - $1,650
Sportster XLCH - $1,600
Electra Glide FLB - $1,735
Electra Glide FLFB - $1,735
Electra Glide FLHB - $1,800

Electra Glide FLHFB - $1,800
Servicar - $1,930
Production
Sportster - 4,500
Electra Glide - 7,750
Servicar - 600
TOTAL - 12,850
Colours
Electra Glide - Black, Crystal Blue, Sparkling Burgundy, Hi-Fi Blue
Servicar - Holiday Red, Birch White
Specifications
Sprint SS - Replaces Sprint (only minor changes), Sprint H continues
XL Sportster - Electric start, revised shock absorbers
FL Electra Glides - No major changes
Servicar - Fibreglass box

1968

The Range
68MS 50 Sport - 3ci (49cc) two-stroke single, three-speed
68M 65 - 4ci (**65cc**) two-stroke single, three-speed
68MS 65 Sport - 4ci two-stroke single, three-speed
68M125 Rapido - 7ci (124cc) two-stroke single, four-speed
68H Sprint H - 15ci (**246cc**) OHV single, four-speed
68SS Sprint SS - 15ci OHV single, four-speed
68XLH Sportster - 54ci (**883cc**) OHV V-twin, four-speed
68XLCH Sportster - 54ci OHV V-twin, four-speed
68FLB Electra Glide Sport - 74ci (**1200cc**) OHV V-twin, four-speed
68FLFB Electra Glide Sport - 74ci OHV V-twin, four-speed, foot-change
68FLHB Electra Glide Super Sport - 74ci OHV V-twin, high-compression, four-speed
68FLHFB Electra Glide Super Sport - 74ci OHV V-twin, high-compression, four-speed, foot-change
68GE Servicar - 45ci (750cc) flathead V-twin, three-speed+reverse
Prices
Sportster XLH - $1,650
Sportster XLCH - $1,600
Electra Glide FLB - $1,735
Electra Glide FLFB - $1,735
Electra Glide FLHB - $1,800
Electra Glide FLHFB - $1,800
Servicar - $1,930
Production
Sportster - 6,875
Electra Glide - 6,950
Servicar - 600
TOTAL - 14,425
Colours
Electra Glide - Black, Jet Fire Orange, Sparkling Burgundy, Hi-Fi Sparkling Blue
Servicar - Birch White
Specifications
XL Sportster - Longer travel front forks
FL Electra Glides - Redesigned wet clutch
Servicar - No major changes

1969

The Range
69M 65 - 3ci (**65cc**) two-stroke single, three-speed
65MS 65 Sport - 3ci two-stroke single, three-speed
69ML 125 Rapido - 7ci (124cc) two-stroke single, three-speed
69SS Sprint SS - 15ci (**246cc**) OHV single, four-speed
69ERS Sprint Scrambler - 15ci single, four-speed
69XLH Sportster - 54ci (**883cc**) OHV V-twin, four-speed
69XLCH Sportster - 54ci OHV V-twin, four-speed
69FLB Super Sport - 74ci (**1200cc**) OHV V-twin, four-speed
69FLFB Super Sport - 74ci OHV V-twin, four-speed, foot-change
69FLHB Super Sport - 74ci OHV V-twin, high-compression, four-speed
69FLHFB Super Sport - 74ci OHV V-twin, high-compression, four-speed, foot-change
69GE Servicar - 45ci (750cc) flathead V-twin, three-speed+reverse
Prices
Sportster XLH - $1,765
Sportster XLCH - $1,698
Electra Glide FLB - $1,885
Electra Glide FLFB - $1,885
Electra Glide FLHB - $1,900
Electra Glide FLHFB - $1,900
Servicar - $2,065
Production
Sportster - 7,800
Electra Glide - 7,300
Servicar - 475
TOTAL - 15,575

The Sportster XLCH of 1968, the cheapest twin in the range, and actually listed at a slightly lower price than the basic XLH. It had higher compression but less equipment, though the XLCH, far from being the 'competition' model it was originally intended to be, would soon be no more than the standard XL with different styling.

Colours
Sportster/Electra Glide - Black, Jet Fire Orange, Birch White, Sparkling Burgundy, Sparkling Blue, Sparkling Gold
Servicar - Birch White
Specifications
Sprint Scrambler - Replaces Sprint H, Sprint SS continues
XL Sportster - Stacked silencers
FL Electra Glides - No major changes
Servicar - No major changes

1970

The Range
70MS 65 Sport Leggero - 4ci (65cc) two-stroke single, three-speed
70MSR Baja - 6ci (98cc) two-stroke single, four-speed
70MLS Rapido - 7ci (124cc) two-stroke single, four-speed
70SS Sprint - 21ci (**346cc**) OHV single, four-speed
70ERS Sprint Scrambler - 21ci OHV single, four-speed
70XLH Sportster - 54ci (**883cc**) OHV V-twin, four-speed
70XLCH Sportster - 54ci OHV V-twin, four-speed
70FLP Super Sport - 74ci (**1200cc**) OHV V-twin, four-speed
70FLPF Super Sport - 74ci OHV V-twin,, four-speed, foot-change
70FLH Super Sport - 74ci OHV V-twin, high-compression, four-speed
70FLHF Super Sport - 74ci OHV V-twin, high-compression, four-speed, foot-change
70GE Servicar - 45ci (750cc) flathead V-twin, three-speed+reverse
Production
Sportster - 8,560
Electra-Glide - 7,615
Servicar - 494
TOTAL - 16,669
Colours
All models - Birch White, Sparkling Blue

LEFT: The 1966/67 Electra Glide marks the debut of the Shovelhead motor, with the familiar Panhead bottom end and capacity, with new cylinder heads. For many buyers, this was just a taste of their own personal dresser.

BELOW LEFT: At the other end of the range, the 7-ci (125-cc) Rapido was an import from Aermacchi in 1969. It was more up to date than the Hummer/Bobcat it replaced, but was still not as sophisticated as the small Japanese bikes.

Specifications
Baja 100 - New model from Aermacchi, 6ci two-stroke single, petroil lubrication, pure off-road machine
MLS Rapido - Now has off-road styling
Sprint SS/Scrambler - New model, enlarged 21ci OHV single replaces 15ci
XL Sportster - Optional 'boat tail' seat/mudguard unit, XLCH has XL's points/coil ignition
FL Electra Glides - Alternator replaces generator, redesigned crankcases
Servicar - No major changes

1971

The Range
71MS 65 Sport Leggero - 4ci (65cc) two-stroke single, three-speed
71MSR Baja - 6ci (98cc) two-stroke single, four-speed

LEFT: Diversification: in the 1970s, Harley-Davidson's search for new markets led it to experiment with snowmobiles; but they failed to stand the test of time.

RIGHT: Meanwhile, the traditional touring market provided far more reliable bread-and-butter income. FLT seen here didn't arrive until 1980, by which time the touring sector had already shrunk to just one piece of a much bigger motorcycle market.

71MLS Rapido - 7ci (124cc) two-stroke single, four-speed
71SS Sprint SS - 21ci (**346cc**) OHV single, four-speed
71SX Sprint SX - 21ci OHV single, four-speed
71XLH Sportster - 54ci (**883cc**) OHV V-twin, four-speed
71XLCH Sportster - 54ci OHV V-twin, four-speed
71FX Super Glide - 74ci (**1200cc**) OHV V-twin, four-speed
71FLP Super Sport - 74ci OHV V-twin, four-speed
71FLPF Super Sport - 74ci OHV V-twin, four-speed, foot-change
71FLH Super Sport - 74ci OHV V-twin, high-compression, four-speed
71FLHF Super Sport - 74ci OHV V-twin, high-compression, four-speed, foot-change
71GE Servicar - 45ci (750cc) flathead V-twin, three-speed+reverse
Production
Sportster - 10,775
Super Glide - 4,700
FL/FLH - 6,675
Servicar - 500
TOTAL - 22,650
Colours
All models - Birch White, Sparkling Blue
Specifications
Sprint SX - Replaces Sprint Scrambler
XL Sportster - Wet clutch with single spring
FX Super Glide - New model, combining FLH 74ci OHV V-twin (but without electric start), frame and running gear with Sportster XLH front end, boat tail rear end, 16in (406mm) front wheel, 19in (483mm) rear, 3.5-gal (16-litre) dual fuel tank
FL/FLH - Bendix/Zenith carburettor, 10in (254mm) front disc brake
Servicar - No major changes

1972

The Range
72MS 65 Sport Leggero - 4ci (**65cc**) two-stroke single, three-speed
72MC 65 Shortster - 4ci two-stroke single, three-speed
72MSR Baja - 6ci (98cc) two-stroke single, four-speed
72MLS Rapido - 7ci (124cc) two-stroke single, four-speed
72SX Sprint SX - 21ci (346cc) OHV single, four-speed
72XLH Sportster - 61ci (**1000cc**) OHV V-twin, four-speed
72XLCH Sportster - 61ci OHV V-twin, four-speed
72FX Super Glide - 74ci (**1200cc**) OHV V-twin, four-speed
72FLP Super Sport - 74ci OHV V-twin, four-speed
72FLPF Super Sport - 74ci OHV V-twin, four-speed, foot-change
72FLH Super Sport - 74ci OHV V-twin, high-compression, four-speed
72FLHF Super Sport - 74ci OHV V-twin, high-compression, four-speed, foot-change
72GE Servicar - 45ci (750cc) flathead V-twin, three-speed+reverse
Production
Sportster - 18,150
Super Glide - 6,500
FL/FLH - 9,700
Servicar - 400
TOTAL - 34,750
Colours
All models - Birch White, Sparkling Blue
Specifications
Shortster 65 - New model, based on M65 with mini-bike styling.
Baja 100 - Lighting option (model Baja 100L)
XL Sportster - Bigger 61ci V-twin, bored-out version of 54ci
FX Super Glide - Boat tail dropped, oil pump updated
FL/FLH - No major changes
Servicar - No major changes

1973

The Range
73Z 90 - 5.5ci (**90cc**) two-stroke single
73X 90 - 5.5ci two-stroke single
73SR 100 - 6ci (98cc) two-stroke single, four-speed
73TX125 - 7ci (124cc) two-stroke single, four-speed
73SS Sprint SS - 21ci (**346cc**) OHV single, four-speed
73SX Sprint SX - 21ci OHV single, four-speed

Three years of Super Glides, the bike that did so much to prop up Harley through the difficult years. It was a clever combination of existing parts to create what looked like a brand-new bike. The 1971 (far left, top) and 1972 (far left, bottom) versions sport the original 'boat tail' rear end, which certainly looked different but horrified the traditionalists. Most of the glassfibre rears were junked in favour of conventional mudguards, as a result of which they are now collectors' items. What most buyers wanted was something more like the 1973 version (left) with a more recognizably Harley rear end. The solo seat is non-standard on this one, as are the twin front disc brakes, though the Super Glide did get a single front disc at the factory that year.

73XLH Sportster - 61ci (**1000cc**) OHV V-twin, four-speed
73XLCH Sportster - 61ci OHV V-twin, four-speed
73FX Super Glide - 74ci (**1200cc**) OHV V-twin, four-speed
73FL - 74ci OHV V-twin, four-speed
73FL - 74ci OHV V-twin, high-compression, four-speed
73GE Servicar - 45ci (750cc) flathead V-twin, three-speed+reverse

Production
Sportster - 20,700
Super Glide - 7,625
FL/FLH - 8,775
Servicar - 425
TOTAL - 37,525

Colours
All models - Birch White

Specifications
SR100 - Now has automatic lubrication
TX125 - Replaces Rapido, similar specification with trail-bike styling
Sprint SS/SX - Electric start
XL Sportster - Disc brakes front and rear, Kayaba front forks
FX Super Glide - Disc brakes front and rear, firmer suspension
FL/FLH - Rear disc brake
Servicar - Rear disc brake

1974

The Range
74Z 90 - 5.5ci (**90cc**) two-stroke single
74X 90 - 5.5ci two-stroke single
74SR100 - 6ci (98cc) two-stroke single, four-speed
74SX 125 - 7ci (124cc) two-stroke single, four-speed
74SX 175 - 11ci (174cc) two-stroke single, five-speed
74SS Sprint SS - 21ci (**346cc**) OHV single, four-speed
74SX Sprint SX - 21 ci OHV single, four-speed
74XLH Sportster - 61ci (**1000cc**) OHV V-twin, four-speed
74XLCH Sportster - 61ci OHV V-twin, four-speed
74FX Super Glide - 74ci (**1200cc**) OHV V-twin, four-speed
74FXE Super Glide - 74ci OHV V-twin, four-speed
74FL Police - 74ci OHV V-twin, four-speed
74FLH - 74ci OHV V-twin, four-speed
74FLHF - 74ci OHV V-twin, four-speed

Production
Sportster - 23,830
Super Glide - 9,233
FLH - 5,166
FLHF - 1,310
FL Police - 791
XR750 - 100
TOTAL - 40,430

Colours
All models - Birch White, Vivid Black, Sunburst Burgundy, Sunburst Blue

Specifications
SX125 - Replaces TX125
SX175 - New model from Aermacchi, similar specification to Yamaha DT-1 trail bike
XL Sportster - Return spring on throttle
FX Super Glide - FXE Super Glide has electric start, with system from FLH. Higher first gear, two-cable throttle, other minor changes
FL/FLH - Keihin carburettors, alarm system

1975

The Range
75Z 90 - 5.5ci (**90cc**) two-stroke single
75X 90 - 5.5ci two-stroke single
75SX 125 - 7ci (**124cc**) two-stroke single, four-speed
74RC 125 - 7ci two-stroke single, four-speed
75SX 175 - 11ci (174cc) two-stroke single, five-speed
75SS 250SS - 15ci (**242cc**) two-stroke single, five-speed
75SX 250SX - 15ci two-stroke single, five-speed
75XLH Sportster - 61ci (**1000cc**) OHV V-twin, four-speed
75XLCH Sportster - 61ci OHV V-twin, four-speed
75FX Super Glide - 74ci (**1200cc**) OHV V-twin, four-speed
75FXE Super Glide - 74ci OHV V-twin, four-speed
75FL Police - 74ci OHV V-twin, four-speed
75FLH - 74ci OHV V-twin, four-speed
75FLHF - 74ci OHV V-twin, four-speed

Production
Sportster - 19,410
Super Glide - 12,410
FLHF - 1,535
FL Police - 900
TOTAL - 34,255

Colours
All models - Birch White, Vivid Black, Sunburst Burgundy, Sunburst Blue, Sunburst/Vivid Orange

Specifications
250SS/SX - New model from Aermacchi, 15ci two-stroke single, five-speed, automatic lubrication, indicators, SS road bike, SX trail bike
XL Sportster - Left-foot gearchange/right-foot brake pedal
FL/FLH - Two-cable throttle

The short-lived Z90 was one of the smaller Aermacchi imports, this one with unshrouded forks and high-level silencer. The high-level pipe was there to imply off-road ability for, at the time, trail bikes were the latest big thing, and were a useful means of giving small machines extra sales appeal in the big bike-centred US market. The Z90 was only listed for one year, however, and within three years the whole Aermacchi liaison had been brought to an end. It would be over 20 years before Harley-Davidson sold another sub-54ci (883cc) machine.

1976

The Range
76SS 125 - 7ci (**124cc**) two-stroke single, four-speed
76SXT 125 - 7ci two-stroke single, four-speed
76SS 175 - 11ci (174cc) two-stroke single, five-speed
76SS 250 - 15ci (**242cc**) two-stroke single, five-speed
76SX 250 - 15ci two-stroke single, five-speed
76XLH Sportster - 61ci (**1000cc**) OHV V-twin, four-speed
76XLCH Sportster - 61ci OHV V-twin, four-speed
76FX Super Glide - 74ci (**1200cc**) OHV V-twin, four-speed
76FXE Super Glide - 74ci OHV V-twin, four-speed
76FLH-1200 - 74ci OHV V-twin, four-speed

Production
Sportster - 18,082
Super Glide - 17,695
FLH-1200 - 11,891
TOTAL - 47,668

Colours
All models - Birch White, Vivid Black, Sunburst Burgundy, Sunburst/Vivid Orange, Vivid Blue, Vivid Red, Vivid Brown, Champagne Silver

Specifications
Liberty Editions of all models, to celebrate USA's bicentennary. Black metalflake main colour, with red, white and blue detailing

1977

The Range
77SS 125 - 7ci (**124cc**) two-stroke single, four-speed
77SXT 125 - 7ci two-stroke single, four-speed
77SS 175 - 11ci (174cc) two-stroke single, five-speed
77SS 250 - 15ci (**242cc**) two-stroke single, five-speed
77SX 250 - 15ci two-stroke single, five-speed
77XLH Sportster - 61ci (**1000cc**) OHV V-twin, four-speed
77XLCH Sportster - 61ci OHV V-twin, four-speed
77XLT Sportster - 61ci OHV V-twin, four-speed
77XLCR Café Racer - 61ci OHV V-twin, four-speed
77FX Super Glide - 74ci (**1200cc**) OHV V-twin, four-speed

77FXE Super Glide - 74ci OHV V-twin, four-speed
77FXS Low Rider - 74ci OHV V-twin, four-speed
77FLH-1200 Electra Glide - 74ci OHV V-twin, four-speed
77FLHS Electra Glide - 74ci OHV V-twin, four-speed

Production
Sportster - 17,915
Café Racer - 1,923
Super Glide - 11,449
FLH/FLHS - 9,226
Low Rider - 3,742
TOTAL - 44,255

Colours
All models - Birch White, Vivid Black, Sunburst Burgundy, Sunburst/Vivid Orange, Vivid Blue, Vivid Red, Vivid Brown, Charcoal Silver, Midnight Blue Metallic, Arctic White

Specifications
XLT Sportster - New model, touring version of XLH, with thicker seat, 3.5-gal (16-litre) tank, touring handlebars, Electra Glide-type saddle bags, windscreen, higher gearing
XLCR Café Racer - New model, combining Sportster power train with redesigned European-style chassis and looks. Redesigned frame, cast aluminium wheels, wrinkle-black finish on engine, low bars, twin front discs, bikini fairing, rearset footpegs
FX/FXE Super Glides - Showa front forks, cast aluminium rear wheel optional
FXS Low Rider - New model, restyled version of FXE Super Glide. Lower frame, raked forks, deeply dished king & queen seat, 3.5-gal (16-litre) dual tanks (instruments on tank console), shorter rear shocks, cast aluminium wheels (19in /483mm) front, 16in/406mm rear), flat drag-style handlebars with risers, highway (set forward) pegs
FLH-1200 - Revised gearbox
FLHS Electra Glide Sports - limited edition of FLH, brown with tan trim

1978

The Range
78SX 250 - 15ci (242cc) two-stroke single, five-speed
78XLH Sportster - 61ci (**1000cc**) OHV V-twin, four-speed
78XLCH Sportster - 61ci OHV V-twin, four-speed
78XLT Sportster - 61ci OHV V-twin, four-speed
78XLCR Café Racer - 61ci OHV V-twin, four-speed

ABOVE LEFT: The RR250 racer was really an Aermacchi, but badged as a Harley-Davidson brought some high-profile racing success to Milwaukee. This 15-ci (250-cc) water-cooled twin won the 250 World Championship three times, from 1974–76, with Walter Villa riding. But it was a last blaze of glory for Aermacchi, which was sold to Cagiva soon afterwards.

ABOVE: The SX250 of 1976, was the final incarnation of the two-stroke Aermacchis with Harley badges. It was fast, and on paper looked the part to meet the Japanese, with a five-speed gearbox and automatic lubrication. But it cost too much, and build quality wasn't up to Honda standards. From 1979, Harley-Davidson accepted that it could (or would have to) survive without making small bikes.

78FX Super Glide - 74ci (**1200cc**) OHV V-twin, four-speed
78FXE Super Glide - 74ci OHV V-twin, four-speed
78FXS Low Rider - 74ci OHV V-twin, four-speed
78FLH-1200 Electra Glide - 74ci OHV V-twin, four-speed
78FLH-80 Electra Glide - 82ci (1340cc) OHV V-twin, four-speed

Sample Prices
XLCH Sportster - $3,370
FLH-80 Electra Glide - $4,905

Production
Sportster - 16,360
Café Racer - 1,201
Super Glide - 10,088
FLH-1200/80 -9,406
Low Rider - 9,787

TOTAL - 46,842

Colours
All models - Birch White, Vivid Black, Vivid Red, Chestnut Brown, Charcoal Silver, Arctic White, Black Cherry

Specifications
XL/XLCH/XLT Sportsters - Twin front discs, electronic ignition, siamezed exhaust system, cast wheels optional, 75th Anniversary version of XLH (Black & Gold with leather seat)
XLCR Café Racer - Dual seat optional
FX/FXE Super Glides - Electronic ignition, cast-iron valve guides, revised intake manifold
FXS Low Rider - Electronic ignition, cast-iron valve guides, revised intake manifold
FLH-1200 - Electronic ignition, cast-iron valve guides, revised cam and intake manifold, cast wheels optional, 75th Anniversary version (Black & Gold with leather seat)
FLH-80 - New 82ci V-twin, bored and stroked version of 74ci twin

The 1978 XLH Sportster, a British-registered bike. Harley-Davidson had reason to be grateful to the Sportster: throughout the 1970s, it was the best-selling bike in the range and in the best year (1974) nearly 24,000 were built. It also introduced many riders to Harleys, who went on to buy bigger machines. The exception was the XLCR Café Racer, which was just too radical for most tastes. Gradual updates (disc brakes, bigger capacity etc) kept many interested in the Sportster through the 1970s.

OPPOSITE:
ABOVE LEFT: Tour Glide was an attempt to radically update the standard touring Harley.

ABOVE RIGHT: Wide Glide married a standard low-riding rear end with widely-spaced front forks. This is a 1980 Shovelhead version.

1979

The Range
79XLH Sportster - 61ci (**1000cc**) OHV V-twin, four-speed
79XLCH Sportster - 61ci OHV V-twin, four-speed
79XLS Sportster - 61ci OHV V-twin, four-speed
79FXE-1200 Super Glide - 74ci (**1200cc**) OHV V-twin, four-speed
79FXS-1200 Low Rider - 74ci OHV V-twin, four-speed
79FXS-80 Low Rider - 82ci (**1340cc**) OHV V-twin, four-speed
79FXEF-1200 Fat Bob - 74ci OHV V-twin, four-speed
79FXEF-80 Fat Bob - 82ci OHV V-twin, four-speed
79FLH-1200 Electra Glide - 74ci OHV V-twin, four-speed
79FLH-80 Electra Glide - 82ci OHV V-twin, four-speed
79FLH-1200 Police Electra Glide - 74ci OHV V-twin, four-speed

79FLH-80 Police Electra Glide - 82ci OHV V-twin, four-speed
79FLHC Electra Glide Classic - 82ci OHV V-twin, four-speed

Production
Sportster - 11,789
Café Racer - 9
Super Glide - 9,942
FLH-1200/80 - 11,442
Low Rider - 13,260
TOTAL - 46,442

Colours
All models - Birch White, Vivid Black, Brilliant Red, Chestnut Brown, Concord Blue, Classic Tan, Classic Creme, FXS Silver, Charcoal Grey

Specifications
XL/XLCH Sportsters - Minor changes
XLS Sportster - New model, based on Sportster with Low Rider styling - raked forks, 16-in (406-mm) rear wheel, drag bars, highway pegs etc
XLCR Café Racer - Discontinued
FXE Super Glide - Twin front discs
FXS Low Rider - Minor changes
FXEF1200/80 Fat Bobs - New models based on Super Glide, with twin tanks, bobbed rear mudguard and buckhorn (higher) bars
FLH Electra Glide Classic - New model, with cast wheels and Tour Pak fairing, saddlebags, crashbars and luggage rack, Tan & Cream colour scheme

1980

The Range
80XLH Sportster - 61ci (**1000cc**) OHV V-twin, four-speed
80XLS Sportster - 61ci OHV V-twin, four-speed
80FXE-1200 Super Glide - 74ci (**1200cc**) OHV V-twin, four-speed
80FXS-1200 Low Rider - 74ci OHV V-twin, four-speed
80FXS-80 Low Rider - 82ci (**1340cc**) OHV V-twin, four-speed
80FXEF-80 Fat Bob - 82ci OHV V-twin, four-speed
80FXEF-80 Fat Bob - 82ci OHV V-twin, four-speed
80FXWG Wide Glide - 82ci OHV V-twin, four-speed
80FXB Sturgis - 82ci OHV V-twin, four-speed
80FLH-1200 Electra Glide - 74ci OHV V-twin, four-speed
80FLH-80 Electra Glide - 82ci OHV V-twin, four-speed
80FLHS Electra Glide - 82ci OHV V-twin, four-speed
80FLHC Electra Glide Classic - 82ci OHV V-twin, four-speed
80FLT Tour Glide - 82ci OHV V-twin, five-speed

Sample Prices
XLH Sportster - $5,867
FXWG Wide Glide - $5,683
FLT Tour Glide - $6,013
FXB Sturgis - $5,687

Production
Sportster - 14,747
Super Glide - 7,942
FLH Electra Glide - 7,512
FLT Tour Glide - 4,480
Low Rider - 5,925
FXWG Wide Glide - 6,085
TOTAL - 46,691

Colours
All models - Birch White, Vivid Black, Rich Red, Saddle Brown, Bright Blue, FXS Silver, Charcoal Grey, Candy Red

Specifications
XL Sportsters - Optional Hugger with lower seat (shorter shocks and thinner seat)
FXE Super Glide - No changes
FXS Low Riders - Motorola electronic ignition, 2-into-1 exhaust
FXB Sturgis - New model, based on FXS-80 Low Rider, with belt primary and final drive, 2-in (51-mm) extended forks, oil cooler, highway pegs, cast wheels
FXEF1200/80 Fat Bobs - Electronic ignition, cast or spoke wheels
FLT Tour Glide - New model, using 82ci V-twin, three-point rubber-mounted to isolate vibration, five-speed gearbox, rectangular steel backbone frame, frame-mounted fairing, enclosed rear chain, twin 10-in (254-mm) front discs, single 12-in (305-mm) rear disc, 16-in (406-mm) wheels, separate (not tank-mounted) instruments
FLH Electra Glide Classic - Electronic ignition
FXWG Wide Glide - New model, using 82ci V-twin and four-speed gearbox, extended wide-spaced front forks, 21-in (533-mm) front wheel, wider front axle, twin fuel tanks (5-gal/22.5-litre) rigidly-mounted engines, low stepped seat (26 1/4-in/67-cm), buckhorn handlebars, staggered shorty dual exhaust system, black colour scheme with orange flames on tank

1981

The Range
81XLH Sportster - 61ci (**1000cc**) OHV V-twin, four-speed
81XLS Sportster - 61ci OHV V-twin, four-speed
81FXE Super Glide - 82ci (**1340cc**) OHV V-twin, four-speed
81FXS Low Rider - 82ci OHV V-twin, four-speed
81FXEF Fat Bob - 82ci OHV V-twin, four-speed

81FXWG Wide Glide - 82ci OHV V-twin, four-speed
81FXB Sturgis - 82ci OHV V-twin, four-speed
81FLH Electra Glide - 82ci OHV V-twin, four-speed
81FLHS Electra Glide - 82ci OHV V-twin, four-speed
81FLH Electra Glide Heritage - 82ci OHV V-twin, four-speed
81FLHC Electra Glide Classic - 82ci OHV V-twin, four-speed
81FLT Tour Glide - 82ci OHV V-twin, five-speed
81FLTC Tour Glide - 82ci OHV V-twin, five-speed

Production

Sportster - 10,102
Super Glide - 3,085
Fat Bob - 3,691
FLH Electra Glide - 6,001
FLT Tour Glide - 2,793
Low Rider - 7,223
FXWG Wide Glide - 5,166
FXB Sturgis - 3,543
TOTAL - 41,604

Colours

All models - Birch White, Vivid Black, Classic Tan, Classic Creme, Candy Root Beer, Metallic Blue, Light Red, Dark Carmine, Orange, Olive, Light Charcoal Metallic, Dark Silver

Specifications

XLH Sportster - Shorter front forks, options include cast wheels, 16-in (406-mm) rear wheel, 3.3-gal (15-litre) tank
XLS Sportster - Buckhorn handlebars, shorty dual exhaust system, 3.3-gal tank, options include cast wheels, 16-in rear wheel, 2.2-gal (10-litre) tank
All 82ci V-twins - V-Fire II electronic ignition, lower 7.4:1 compression ratio, longer valve guides, extra oil drain lines from rocker boxes, revised valve guide seals, Girling rear brake caliper

FAR LEFT:
ABOVE: Even in 1981, some people still opted for a sidecar. This is a Heritage edition of the standard FLH.

BELOW: At a glance, this bike could have been made at any time during the last 50 years. However, a closer look at the Shovelhead motor narrows it down to between 1966 and 1982.

LEFT: The 1982 FXB Sturgis brought belt drive to a new Harley for the first time, though unlike present belt-drive Harley-Davidsons, this one had a primary belt as well. It was a big step forward by the engineering department, needing no lubrication or adjustment, helping to smooth out the transmission and keeping the back of the bike cleaner.

FLH Electra Glide Heritage - New model, FLH with retro styling, fringed leather saddlebags, windscreen, suspended seat
FLT Tour Glides - Revised exhaust, FLTC had special colour scheme

1982

The Range
82XLH Sportster - 61ci (**1000cc**) OHV V-twin, four-speed
82XLS Roadster - 61ci OHV V-twin, four-speed
82FXE Super Glide - 82ci (**1340cc**) OHV V-twin, four-speed
82FXS Low Rider - 82ci OHV V-twin, four-speed
82FXWG Wide Glide - 82ci OHV V-twin, four-speed
82FXB Sturgis - 82ci OHV V-twin, four-speed
82FXR Super Glide II - 82ci OHV V-twin, five-speed
82FXRS Super Glide II - 82ci OHV V-twin, five-speed
82FLH Electra Glide - 82ci OHV V-twin, four-speed
82FLT Tour Glide - 82ci OHV V-twin, five-speed
82FLTC Tour Glide - 82ci OHV V-twin, five-speed

Production
Sportster - 5,947
Super Glide - 1,617
Super Glide II - 6,255
FLH Electra Glide - 6,185
FLT Tour Glide - 2,029
Low Rider - 1,816
FXWG Wide Glide - 2,348
FXB Sturgis - 1,833
TOTAL - 28,030

Colours
All models - Birch White, Vivid Black, Classic Tan, Classic Creme, Metallic Blue, Light Red, Dark Carmine, Orange, Brown, Emerald Green, Red, Pewter, Maroon

Specifications
XLH Sportster /XLS Roadster - Lighter frame, lower 8:1 compression ratio
FXB Sturgis - Higher bars, gold cast wheels
FLT Tour Glide - New handlebars, seat, primary chain oiler, higher output alternator, locks for luggage
FXR/FXRS Super Glide II - New model, using 81ci V-twin, rubber-mounted, bar-mounted instruments, new frame, twin 10-in (254-mm) front discs, 19-in (483-mm) front wheel, 16-in (406-mm) rear wheel, wire wheels. FXRS has sissy bar, stepped seat, highway pegs, two-tone paint and cast wheels

ABOVE LEFT and ABOVE
In 1982, the FX family was in full swing. Little did Willie G. realize (or maybe he did) that the original FX Super Glide would spawn a whole new range of Harleys, slotting between the Sportsters on one hand and the FLH/FLT tourers on the other. They made a virtue out of custom styling, of which the Super Glide was a pioneer for factory machines. The 1978 Low Rider was the first (labelled 'the thinking man's chopper' by *Cycle World* magazine), and was an instant hit. Fat Bob, Sturgis and Wide Glide soon followed.

OPPOSITE: The sportiest Harley of its day. The XR1000 was a seriously souped-up XLX, with aluminium XR-type high-compression cylinder heads, twin carburettors, twin disc brakes and cast wheels. It was fast but expensive, and at the time arguably the sort of indulgence Harley could well have done without.

1983

The Range

83XLX Sportster - 61ci (**1000cc**) OHV V-twin, four-speed
83XLH Sportster -61ci OHV V-twin, four-speed
83XLS Roadster - 61ci OHV V-twin, four-speed
83XR1000 Sportster - 61ci OHV V-twin, four-speed
83FXE Super Glide - 82ci (**1340cc**) OHV V-twin, four-speed
83FXWG Wide Glide - 82ci OHV V-twin, four-speed
83FXSB Low Rider - 82ci OHV V-twin, four-speed
83FXDG Disc Glide - 82ci OHV V-twin, four-speed
83FXR Super Glide II - 82ci OHV V-twin, five-speed
83FXRS Super Glide II - 82ci OHV V-twin, five-speed
83FXRT Sport Glide - 82ci OHV V-twin, five speed
83FLH Electra Glide - 82ci OHV V-twin, four-speed
83FLHS Electra Glide Sport - 82ci OHV V-twin, four-speed
83FLHT Electra Glide - 82ci OHV V-twin, five-speed
83 FLHTC Electra Glide Classic - 82ci OHV V-twin, five-speed
83FLT Tour Glide - 82ci OHV V-twin, five-speed
83FLTC Tour Glide Classic - 82ci OHV V-twin, five-speed

Production

Sportster - 9,756
Super Glide - 1,215
Super Glide II - 4,810
FLH Electra Glides - 6,485
FLT Tour Glide - 1,078
Low Rider - 3,277
Wide Glide - 2,873

TOTAL - 29,494

Colours

All models - Birch White, Vivid Black, Classic Tan, Classic Creme, Pewter, Maroon, Indigo Blue, Claret Red, Slate Grey Metallic

Specifications

XLH Sportster - High bars, vacuum-advance ignition, 3.3-gal (15-litre) tank, new seat, less restrictive exhaust system
XLX Sportster - New model, a stripped-down version to sell for $3,995. Lower bars, solo seat, peanut fuel tank, cast wheels, no tachometer, in black only
XR1000 Sportster - New model based on XLX, with all-new top end – aluminium XR-style cylinder heads, ported and polished by tuner Jerry Branch, 9:1 compression ratio, V-Fire III electronic ignition, twin 1.4-in (36-mm) Dell'Orto carburettors, megaphone silencers, cast wheels
FLH Electra Glide - Belt final drive
FLHT/C Electra Glide - New models, combining FLT chassis with FLH-style frame-mounted fairing, 16-in (406-mm) rear wheel, larger luggage
FLT Tour Glides - Lower seat, stiffer suspension
FXDG Disc Glide - New model, based on Wide Glide with aluminium disc rear wheel, belt final drive and black shorty dual exhaust
FXSB Low Rider - New model, based on FXS Low Rider with belt final drive and 21-in (533-mm) front wheel
FXRT Sport Glide - New model based on FXRS chassis, classified as sports-tourer – aerodynamic frame-mounted fairing, anti-dive front forks, enclosed rear chain, adjustable air suspension, frame-mounted panniers, low touring handlebars, adjustable pillion pegs, 4.2-gal (19-litre) tank, various touring options

1984

The Range

84XLX Sportster - 61ci (**1000cc**) OHV V-twin, four-speed
84XLH Sportster - 61ci OHV V-twin, four-speed
84XLS Roadster - 61ci OHV V-twin, four-speed
84XR1000 Sportster - 61ci OHV V-twin, four-speed
84FXE Super Glide - 82ci (**1340cc**) OHV V-twin, four-speed
84FXST Softail - 82ci Evolution V-twin, four-speed
84FXWG Wide Glide - 82ci OHV V-twin, four-speed
84FXSB Low Rider - 82ci twin, four-speed
84FXRDG Disc Glide - 82ci Evolution V-twin, four-speed
84FXRS Low Glide - 82ci Evolution V-twin, five-speed
84FXRT Sport Glide - 82ci Evolution V-twin, five speed
84FXRP Police - 82ci Evolution V-twin, five-speed
84FLH Electra Glide - 82ci OHV V-twin, four-speed
84FLHS Electra Glide Sport - 82ci OHV V-twin, four-speed
84FLHX Electra Glide - 82ci OHV V-twin, four-speed
84FLHTC Electra Glide Classic - 82ci Evolution V-twin, five-speed
84FLTC Tour Glide Classic - 82ci Evolution V-twin, five-speed

Sample Prices

FXRS Low Glide - $7,560
FXDRG Disc Glide - $8,199
FLHTC Electra Glide Classic - $8,799

Production

Sportster - 10,620
Super Glide - 2,107
FXRS/T/DG - 5,703
Softail - 5,413

FLH Electra Glide - 7,569
FLT Tour Glide - 1,334
Low Rider - 2,877
Wide Glide - 2,227
TOTAL - 37,850
Colours
All models - Birch White, Vivid Black, Classic Tan, Classic Creme, Pewter, Maroon, Indigo Blue, Claret Red, Candy Purple, Rose Metallic, Slate Grey Metallic
Specifications
XLH/XLX Sportsters - Single front disc, diaphragm clutch, alternator
XR1000 Sportster - Improved brakes, orange/black racing colours optional
FXRS/T/P - Evo V-twin, diaphragm clutch – FXRS Low Glide has shorter rear shocks and front forks
FXRDG - One-year limited edition based on FXRS with disc rear wheel, different paint and graphics
FXST Softail - New model, based on Wide Glide with hidden rear shocks for hardtail look - belt drive, rigidly-mounted Evo V-twin, four-speed gearbox, triangulated swinging arm, 21-in (533-mm) front wheel, 16-in (406-mm) rear wheel, buckhorn handlebars, wire wheels
FXWG Wide Glide - Belt drive, chain primary, later 1984 models may have used Evo V-twin
FLH/S/X Electra Glide - Diaphragm clutch, staggered dual exhausts, highway pegs
FLHTC Electra Glide - Evo V-twin, diaphragm clutch, air shocks, air front forks with anti-dive, enclosed rear chain
FLTC Tour Glide Classic - Evo V-twin, belt drive, five-speed gearbox, diaphragm clutch

1985

The Range
85XLX Sportster - 61ci (**1000cc**) OHV V-twin, four-speed
85XLH Sportster - 61ci OHV V-twin, four-speed
85XLS Roadster - 61ci OHV V-twin, four-speed
85FXEF Fat Bob - 82ci (**1340cc**) Evo V-twin, five-speed
85FXST Softail - 82ci Evo V-twin, five-speed
85FXWG Wide Glide - 82ci Evo V-twin, four-speed
85FXSB Low Rider - 82ci Evo V-twin, four-speed
85FXRS Low Glide - 82ci Evo V-twin, five-speed
85FXRT Sport Glide - 82ci Evo V-twin, five-speed
85FXRC Low Glide Custom - 82ci Evo V-twin, five-speed
85FLHTC Electra Glide Classic - 82ci Evo V-twin, five-speed
85FLTC Tour Glide Classic - 82ci Evo V-twin, five-speed
Sample Prices
XLX Sportster - $4,695
FXST Softail - $8,399
Production
Sportster - 6,514
FXRS/T/DG - 8,095
Softail - 4,529
FLH Electra Glide - 4,793
FLT Tour Glide - 1,847
Low Rider - 2,359
Wide Glide - 4,171
TOTAL - 32,308
Colours
All models - Birch White, Vivid Black, Classic Tan, Classic Creme, Candy Blue, Slate Grey Metallic
Specifications
XLH/XLX Sportsters - No changes
FXRS/T - Belt final drive
FXRC Low Glide Custom - One-year limited edition based on FXRS, chrome engine covers, wire wheels, Candy Orange
FXEF Fat Bob – Evo V-twin, wet-multiplate clutch, belt final drive
FXSB Low Rider - Evo V-twin, wet multiplate clutch, belt final drive
FXWG Wide Glide - Evo V-twin, wet multiplate clutch, belt final drive

1986

The Range
86XLH 883 Sportster 54ci (883cc) Evo V-twin, four-speed
86XLH 1100 Sportster - 67ci (1100cc) Evo V-twin, four-speed
86FLST Heritage Softail - 82ci (**1340cc**) Evo V-twin, five-speed
86FXR Super Glide - 82ci Evo V-twin, five-speed
86FXRS Low Glide - 82ci Evo V-twin, five-speed
86FXRT Sport Glide - 82ci Evo V-twin, five-speed
86FXRD Sport Glide Grand Touring - 82ci Evo V-twin, five-speed
86FXST Softail - 82ci Evo V-twin, five-speed
86FXSTC Softail Custom - 82ci Evo V-twin, five-speed
86FXWG Wide Glide - 82ci Evo V-twin, four-speed
86FLHTC Electra Glide Classic - 82ci Evo V-twin, five-speed
86FLTC Tour Glide Classic - 82ci Evo V-twin, five-speed
Sample Prices
XLH 883 Sportster - $3,995
XLH 1100 Sportster - $5,199
FLHTC Electra Glide Classic - $10,224
FLTC Tour Glide Classic & sidecar - $13,423
Production
Sportsters - 14,379
FXR/S/T/D - 8,680
Softails - 8,694
FLH Electra Glide - 4,456
FLT Tour Glide - 1,442
Wide Glide - 1,199
TOTAL - 38,850

OPPOSITE: 1985 Electra Glide Classic with full touring kit.

ABOVE: The 1987 Softail Custom used the now ubiquitous Evolution 82-ci (1340-cc) V-twin, as did the whole range by then. The Evo was a hugely important engine for Harley-Davidson, originally intended as a stopgap before an all-new range of water-cooled motors was ready. It was only a little more powerful than the Shovelhead, but was more reliable and better built, with closer tolerances. This engine carried the company through the next 15 years, until replaced by the Twin Cam in 1998. The bike illustrated is also significant in being a Softail. It was a master stroke of marketing to launch this hardtail-look rear end just as nostalgia was becoming big business. This popular model promised traditional looks with at least some modern comfort, though suspension travel was limited compared to a conventional rear suspension.

Colours
All models - Birch White, Vivid Black, Candy Red, Slate Grey Metallic, Candy Blue, Candy Pearl, Signal Red

Specifications
XLH 883 Sportster - Major update of XLH Sportster, with smaller 54ci version of Evo V-twin (3 x 3.81in/76 x 97mm) bore x stroke), aluminium heads and cylinders, 1.3-in (33-mm) Keihin carburettor, four-speed gearbox, new frame, cast wheels (19in/483mm) front, 16in/406mm rear), low bars, solo seat, peanut tank – upgrade package with different paint and trim
XLH 1100 Sportster - As 883 Sportster with 67ci (1100cc) Evo (3.35 x 3.8in/85.1 x 96.5mm bore x stroke)
FXR Super Glide - New model, stripped-down version of FXRS
FXRS Low Glide - Revised indicator switch, optional Sport package with longer-travel suspension and twin front discs
FXRD Sport Glide Grand Touring - Based on FXRT with air rear suspension, rubber-mounted bars, high-spec stereo, full luggage, 2-into-1 exhaust
FXST Softail - Five-speed gearbox
FLST Heritage Softail - New model, based on Softail with fork shrouds, 16-in (406-mm) wire wheels and circular air cleaner; aims to imitate style of 1949–57 Hydra Glide

1987

The Range
87XLH 883 Sportster - 54ci (883cc) Evo V-twin, four-speed
87XLH 1100 Sportster - 67ci (1100cc) Evo V-twin, four-speed
87FLST Heritage Softail - 82ci (**1340cc**) Evo V-twin, five-speed
87FLSTC Heritage Softail Special - 82ci Evo V-twin, five-speed
87FXR Super Glide - 82ci Evo V-twin, five-speed
87FXRS Low Rider - 82ci Evo V-twin, five-speed
87FXRS-SP Low Rider Sport - 82ci Evo V-twin, five-speed
87FXLR Low Rider Custom - 82ci Evo V-twin, five-speed
87FXRT Sport Glide - 82ci Evo V-twin, five-speed
87FXST Softail - 82ci Evo V-twin, five-speed
87FXSTC Softail Custom - 82ci Evo V-twin, five-speed
87FLHS Electra Glide Sport - 82ci Evo V-twin, five-speed
87FLHTC Electra Glide Classic - 82ci Evo V-twin, five-speed
87FLTC Tour Glide Classic - 82ci Evo V-twin, five-speed

Sample Prices
FXRS Low Rider - $8,449
FXSTC Softail Custom - $9,499
FLHTC Electra Glide Classic - $10,395

Production
Sportsters - 13,794
FXR/S/T - 8,000
Softails - 7,686
FLH Electra Glide - 6,452
FLT Tour Glide - 225
TOTAL - 36,157
Specifications
FLHS Electra Glide Sport - Same spec as FLHTC, detachable screen, no fairing, top box or sissy bar
FXR Super Glide - Belt final drive
FXRT Sport Glide - Improved stereo
FLST Heritage Softail - Special package includes windscreen, passing lamps, leather saddlebags, two-piece seat

1988

The Range
88XLH 883 Sportster - 54ci (**883cc**) Evo V-twin, four-speed
88XLH 883 Sportster Hugger - 54ci Evo V-twin, four-speed
88XLH 883 Sportster Deluxe - 54ci Evo V-twin, four-speed
88XLH 1200 Sportster - 73ci (1203cc) Evo V-twin, four-speed
88FLST Heritage Softail - 82ci (**1340cc**) Evo V-twin, five-speed
88FLSTC Heritage Softail Special - 82ci Evo V-twin, five-speed
88FXR Super Glide - 82ci Evo V-twin, five-speed
88FXRS Low Rider - 82ci Evo V-twin, five-speed
88FXRS-SP Low Rider Sport - 82ci Evo V-twin, five-speed
88FXLR Low Rider Custom - 82ci Evo V-twin, five-speed
88FXRT Sport Glide - 82ci Evo V-twin, five-speed
88FXST Softail - 82ci Evo V-twin, five-speed
88FXSTC Softail Custom - 82ci Evo V-twin, five-speed
88FXSTS Springer Softail - 82ci Evo V-twin, five-speed
88FLHS Electra Glide Sport - 82ci Evo V-twin, five-speed
88FLHTC Electra Glide Classic - 82ci Evo V-twin, five-speed
88FLTC Tour Glide Classic - 82ci Evo V-twin, five-speed
Sample Prices
XLH1200 Sportster - $5,875
FXST Softail - $9,375
FLHTC Electra Glide Classic - $10,545
Production
Sportsters - 16,533
FXR/S/T - 7,119
Softails - 15,408
FLH Electra Glides - 7,314
FLT Tour Glides - 848
TOTAL - 47,222
Colours
Single-tones: Vivid Black, Bright Cobalt Candy Blue, Bright Candy Plum, Candy Brandywine, Candy Crimson,
Two-tones: Brandywine & Crimson, Bright Cobalt Blue & Brilliant Silver, Candy Bronze & Creme, Vivid Black & Creme, Metallic Blue & Cadet Blue Metallic, Creme & Champagne Gold

ABOVE and RIGHT: More examples of the late 1980s Softail line. The standard Softail (1987, above) and Softail Custom (1988, below) were actually very similar, but the range was soon expanded to include the Springer Softail as well. Both these bikes are pictured in typical rally scenes.

Specifications
XLH883 Sportsters - 1.6-in (41-mm) Keihin carburettor, longer rear shocks, longer swinging arm, 1.5-in (38-mm) front forks, solo seat, cast wheels. Hugger package includes lower seat and buckhorn bars. Deluxe package includes two-tone paint, buckhorn bars, wire wheels, dual seat, extra chrome, tachometer
XLH1200 Sportster - Replaces XLH1100 with 73ci Evo V-twin (3.498 x 3.812in/88.8 x 96.8mm bore x stroke), 1.5-in (38-mm) front forks, otherwise as 883 Deluxe
FXR/S/LR/RT/ST - 39-mm front forks
FXSTS Springer Softail - New model, based on FXST with exposed springer front forks, much updated version of original 1903 spring fork

1989

The Range
89XLH 883 Sportster - 54ci (**883cc**) Evo V-twin, four-speed
89XLH 883 Sportster Hugger - 54ci Evo V-twin, four-speed
89XLH 883 Sportster Deluxe - 54ci Evo V-twin, four-speed
89XLH 1200 Sportster - 73ci (1203cc) Evo V-twin, four-speed
89FLST Heritage Softail - 82ci (**1340cc**) Evo V-twin, five-speed
89FLSTC Heritage Softail Special - 82ci Evo V-twin, five-speed
89FXR Super Glide - 82ci Evo V-twin, five-speed
89FXRS-CONV Low Rider Convertible - 82ci Evo V-twin, five-speed
89FXRS-SP Low Rider Sport - 82ci Evo V-twin, five-speed
89FXLR Low Rider Custom - 82ci Evo V-twin, five-speed
89FXRT Sport Glide - 82ci Evo V-twin, five-speed
89FXST Softail - 82ci Evo V-twin, five-speed
89FXSTC Softail Custom - 82ci Evo V-twin, five-speed
89FXSTS Springer Softail - 82ci Evo V-twin, five-speed
89FLHS Electra Glide Sport - 82ci Evo V-twin, five-speed
89FLHTC Electra Glide Classic - 82ci Evo V-twin, five-speed
89FLHTCU Electra Glide Classic Ultra - 82ci Evo V-twin, five-speed
89FLTC Tour Glide Classic - 82ci Evo V-twin, five-speed
89FLTCU Tour Glide Classic Ultra - 82ci Evo V-twin, five-speed
Sample Prices
FXRS Convertible - $9,475
FXSTS Springer Softail - $10,759
Production
Sportsters - 16,967
FXR/S/T/P/CONV - 7,015
Softails - 19,756
FLH Electra Glides - 10,148
FLT Tour Glides - 1,185
TOTAL - 55,071
Colours
Single tones: Vivid Black, Metallic Blue, Candy Brandywine, Bright Candy Plum,
Two-tones: Brandywine & Crimson, Metallic Blue & Brilliant Silver, Vivid Black & Creme, Candy Bronze & Creme, Metallic Blue & Cadet Blue Metallic, Creme & Champagne Gold, Vivid Black & Brilliant Silver
Specifications
XLH883/1200 Sportsters - Aluminium intake manifold, offset gudgeon pins, new air cleaner
FXR/S/T/P - New starter, one-piece pinion shaft and right flywheel
FXRS-CONV - New model, based on FXRS with quickly-detachable screen and leather saddlebags, highway pegs and sissy bar
FLST Heritage Softail - 32-amp alternator, one-piece pinion shaft and right flywheel
FLHS/T Electra Glides - 32-amp alternator, one-piece pinion shaft and right flywheel
FLHTCU Electra Glide Classic Ultra - As FLHTC with cruise control, CB radio, intercom, front and rear sound systems, fairing lowers
FLTC Tour Glide - 32-amp alternator, one-piece pinion shaft and right flywheel
FLTCU Tour Glide Classic Ultra - As FLTC with cruise control, CB radio, intercom, front and rear sound systems, fairing lowers

1990

The Range
90XLH 883 Sportster - 54ci (**883cc**) Evo V-twin, four-speed
90XLH 883 Sportster Hugger - 54ci Evo V-twin, four-speed
90XLH 883 Sportster Deluxe - 54ci Evo V-twin, four-speed
90XLH 1200 Sportster - 73ci (1203cc) Evo V-twin, four-speed
90FLST Heritage Softail - 82ci (**1340cc**) Evo V-twin, five-speed
90FLSTC Heritage Softail Classic - 82ci Evo V-twin, five-speed
90FLSTF Fat Boy - 82ci Evo V-twin, five-speed
90FXR Super Glide - 82ci Evo V-twin, five-speed
90FXRS Low Rider - 82ci Evo V-twin, five-speed
90FXRS-CONV Low Rider Convertible - 82ci Evo V-twin, five-speed
90FXRS-SP Low Rider Sport - 82ci Evo V-twin, five-speed
90FXLR Low Rider Custom - 82ci Evo V-twin, five-speed
90FXRT Sport Glide 82ci Evo V-twin, five-speed
90FXST Softail - 82ci Evo V-twin, five-speed
90FXSTC Softail Custom - 82ci Evo V-twin, five-speed
90FXSTS Springer Softail - 82ci Evo V-twin, five-speed
90FLHS Electra Glide Sport - 82ci Evo V-twin, five-speed
90FLHTC Electra Glide Classic - 82ci Evo V-twin, five-speed
90FLHTCU Electra Glide Classic Ultra - 82ci Evo V-twin, five-speed
90FLTC Tour Glide Classic - 82ci Evo V-twin, five-speed
90FLTCU Tour Glide Classic Ultra - 82ci Evo V-twin, five-speed
Sample Prices
XLH883 Sportster - $4,250
FLSTF Fat Boy - $10,995
FLHTCU Electra Glide Classic Ultra - $13,695
Production
Sportsters - 15,163
FXR/S/T/P/CONV - 8,485
Softails - 24,138
FLH Electra Glides - 10,313
FLT Tour Glides - 1,108
TOTAL - 59,207
Colours
Single tones: Vivid Black, Bright Candy Plum, Bright Candy Ruby, Bright Candy Hi-Fi Blue, Fine Silver Metallic, Dark Candy Ruby.
Two-tones: Bright Candy Hi-Fi Blue & Silver, Bright & Dark Candy Ruby, Vivid Black & Black Pearl, Vivid Black & Creme, Creme & Champagne Gold, Bright & Dark Candy Hi-Fi Blue
Specifications
XLH883/1200 Sportsters - 1.57-in (40-mm) Keihin CV carburettor
FXR/S/T/P/CONV - 40-mm Keihin CV carburettor, one-piece right-hand flywheel, redesigned diaphragm clutch
FLSTS Fat Boy - New model, based on FLST Heritage Softail - 16-in (406-mm) disc wheels, wide FLH bars, dual shotgun exhaust system, redesigned clutch, Silver

with Dark Yellow highlights
FXST/FLST - 40-mm Keihin CV carburettor, one-piece right-hand flywheel, redesigned clutch
FLTC/U - 40-mm Keihin CV carburettor, one-piece right hand flywheel, redesigned clutch
FLHTC/U - 40-mm Keihin CV carburettor, one-piece right-hand flywheel, redesigned clutch

1991

The Range
91XLH 883 Sportster - 54ci (**883cc**) Evo V-twin, five-speed
91XLH 883 Sportster Hugger - 54ci Evo V-twin, five-speed
91XLH 883 Sportster Deluxe - 54ci Evo V-twin, five-speed
91XLH 1200 Sportster - 73ci (1203cc) Evo V-twin, five-speed
91FLSTC Heritage Softail Classic - 82ci (**1340cc**) Evo V-twin, five-speed
91FLSTF Fat Boy - 82ci Evo V-twin, five-speed
91FXR Super Glide - 82ci Evo V-twin, five-speed
91FXRS Low Rider - 82ci Evo V-twin, five-speed
91FXRS-CONV Low Rider Convertible - 82ci Evo V-twin, five-speed
91FXRS-SP Low Rider Sport - 82ci Evo V-twin, five-speed
91FXLR Low Rider Custom - 82ci Evo V-twin, five-speed
91FXRT Sport Glide - 82ci Evo V-twin, five-speed
91FXSTC Softail Custom - 82ci Evo V-twin, five-speed
91FXSTS Springer Softail - 82ci Evo V-twin, five-speed
91FXDB Dyna Glide Sturgis - 82ci Evo V-twin, five-speed
91FLHS Electra Glide Sport - 82ci Evo V-twin, five-speed
91FLHTC Electra Glide Classic - 82ci Evo V-twin, five-speed
91FLHTCU Electra Glide Classic Ultra - 82ci Evo V-twin, five-speed
91FLTC Tour Glide Classic - 82ci Evo V-twin, five-speed
91FLTCU Tour Glide Classic Ultra - 82ci Evo V-twin, five-speed

Sample Prices
XLH1200 Sportster - $6,245
FXRS Low Rider - $9,760
FXDB Sturgis - $11,520

Production
Sportsters - 17,725
FXR/S/T/P/CONV/DB - 10,077
Softails - 26,322
FLH Electra Glides - 10,098
FLT Tour Glides - 787
TOTAL - 65,009

Colours
Single tones: Vivid Black, Bright Sapphire Metallic, Wineberry Pearl, Vivid Yellow, Bright Candy Ruby, Candy Sapphire Sun-glo
Two-tones: Bright and Dark Sapphire Metallic, Bright and Dark Candy Ruby, White and Vivid Yellow, White and Turquoise, White and Yellow

Specifications
XLH883 Sportsters - Five-speed gearbox,

ABOVE: A standard Softail from 1990. In the late 1980s, the Softails took over from Sportsters as the best-selling Harleys, which shows just how far the nostalgia market had developed. Over 26,000 were made in 1991.

OPPOSITE: A 1991 Dyna Glide Daytona. The Dyna Glide didn't look very new, but it was. The new steel backbone frame allowed the big V-twin to be rubber-mounted, substantially reducing vibration levels. Otherwise it was the usual canny mix of existing components, with the 82-ci (1340-cc) Evo twin and FXRS front end.

belt drive (Deluxe only), rubber-mounted pegs, self-cancelling indicators, several minor engine changes
XLH1200 Sportster - Five-speed gearbox, belt drive, self-cancelling indicators, several minor engine changes
FXR/S/T/P/CONV - Kevlar base gasket, graphite head gasket, self-cancelling indicators, four-sided fuel inlet valve
FLST/FXST/FLSTF - Kevlar base gaskets, graphite head gaskets, four-sided fuel inlet valve, self-cancelling indicators, other minor changes, black and chrome engine (FLSTC)
FXDB Dyna Glide Sturgis - New model, Evo 82ci V-twin, new frame with rubber mounts using large-diameter rectangular backbone welded to cast steering head, rubber engine mounting system with new engine tilt and location, FXRS forks and front end, single front disc brake, 19-in (483-mm) front wheel, 16-in (406-mm) rear, all engine changes as FX/FL series
FLTC/U – Kevlar base gaskets, graphite head gaskets, four-sided fuel inlet valve
FLHTC/U - Kevlar base gaskets, graphite head gaskets, four-sided fuel inlet valve

1992

The Range
92XLH 883 Sportster - 54ci (**883cc**) Evo V-twin, five-speed
92XLH 883 Sportster Hugger - 54ci Evo V-twin, five-speed
92XLH 883 Sportster Deluxe - 54ci Evo V-twin, five-speed
92XLH 1200 Sportster - 73ci (1203cc) Evo V-twin, five-speed
92FLSTC Heritage Softail Classic - 82ci (**1340cc**) Evo V-twin, five-speed
92FLSTF Fat Boy - 82ci Evo V-twin, five-speed
92FXR Super Glide - 82ci Evo V-twin, five-speed
92FXRS-CONV Low Rider Convertible - 82ci Evo V-twin, five-speed
92FXRS-SP Low Rider Sport - 82ci Evo V-twin, five-speed
92FXLR Low Rider Custom - 82ci Evo V-twin, five-speed
92FXRT Sport Glide - 82ci Evo V-twin, five-speed
92FXSTC Softail Custom - 82ci Evo V-twin, five-speed
92FXSTS Springer Softail - 82ci Evo V-twin, five-speed
92FXDB Dyna Glide Daytona - 82ci Evo V-twin, five-speed
92FXDC Dyna Glide Custom - 82ci Evo Vtwin, five-speed
92FLHS Electra Glide Sport - 82ci Evo V-twin, five-speed
92FLHTC Electra Glide Classic - 82ci Evo V-twin, five-speed

92FLHTCU Electra Glide Classic Ultra - 82ci Evo V-twin, five-speed
92FLTC Tour Glide Classic - 82ci Evo V-twin, five-speed
92FLTCU Tour Glide Classic Ultra - 82ci Evo V-twin, five-speed

Sample Prices
XLH883 Sportster Hugger - $5,075
FLHTC Electra Glide Classic - $15,350
FXDB Daytona - $12,120

Total Production - 76,000

Colours
Vivid Black, True Pearl etc

Specifications
XLH883/1200 Sportsters - New grease fittings, halogen headlight, new horn, other minor changes
FXDB Dyna Glide Daytona - New limited-edition model based on Dyna Glide Sturgis; twin front discs, buckhorn bars, gold wheels, two-tone Indigo Blue Metallic & Gold Pearlglo colour scheme
FXDC Dyna Glide Custom - New model based on Dyna Glide Sturgis; twin front discs, two-tone Silver and Black, silver frame, unpainted engine/gearbox
FXR/S/T/P/CONV - New grease fittings, new horn, caged needle bearing for camshaft
FLST/FXST/FLSTF - New grease fittings, new horn, caged needle bearing for camshaft
FLHS/TC/U - New grease fittings, new horn, caged needle bearing for camshaft, lower seat (FLHS), glossy luggage interior
FLTC/U - New grease fittings, new horn, caged needle bearing for camshaft, glossy luggage interior

1993

The Range
93XLH 883 Sportster - 54ci (**883cc**) Evo V-twin, five-speed
93XLH 883 Sportster Hugger - 54ci Evo V-twin, five-speed
93XLH 883 Sportster Deluxe - 54ci Evo V-twin, five-speed
93XLH 1200 Sportster - 73ci (1203cc) Evo V-twin, five-speed
93FLSTC Heritage Softail Classic- 82ci (**1340cc**) Evo V-twin, five-speed
93FLSTN Heritage Softail Nostalgia - 82ci Evo V-twin, five-speed
93FLSTF Fat Boy - 82ci Evo V-twin, five-speed
93FXR Super Glide - 82ci Evo V-twin, five-speed
93FXRS-CONV Low Rider Convertible - 82ci Evo V-twin, five-speed
93FXRS-SP Low Rider Sport - 82ci Evo V-twin, five-speed
93FXLR Low Rider Custom - 82ci Evo V-twin, five-speed
93FXSTC Softail Custom - 82ci Evo V-

twin, five-speed
93FXSTS Springer Softail - 82ci Evo V-twin, five-speed
93FXDL Dyna Low Rider - 82ci Evo V-twin, five-speed
93FXDWG Dyna Wide Glide - 82ci Evo V-twin, five-speed
93FLHS Electra Glide Sport - 82ci Evo V-twin, five-speed
93FLHTC Electra Glide Classic - 82ci Evo V-twin, five-speed
93FLHTCU Electra Glide Classic Ultra - 82ci Evo V-twin, five-speed
93FLTC Tour Glide Classic - 82ci Evo V-twin, five-speed
93FLTCU Tour Glide Classic Ultra - 82ci Evo V-twin, five-speed

Sample Prices

XLH883 Sportster Deluxe - $5,820
FLSTN Heritage Nostalgia - $13,000
FXDL Dyna Low Rider - $11,800
FLHTCU Electra Glide Classic Ultra (90th Anniversary) - $16,100

Total Production - 82,000

Colours

Single tones: Vivid Black, Mandarin Orange, Scarlet Red, Victory Red Sun-Glo, Bright Aqua Sun-Glo, 90th Anniversary, Bright Wineberry Sun-Glo
Two-tones: Victory Red Sun-Glo & Aqua Sun-Glo, Black & Scarlet Red

Specifications

XLH883 Sportsters - Belt drive on all 883s
FFXR/S/T/P/CONV - No major changes
FXDWG Dyna Wide Glide - New model, using Dyna Glide chassis with wide-set front forks, high bars and 21-in (533-mm) front wheel of earlier Wide Glide, bobbed rear mudguard, single front disc brake, one-piece Fat Bob fuel tank
FXDL Dyna Low Rider - New model using Dyna Glide chassis - removable pillion seat, twin front discs, black and chrome engine/gearbox, staggered shorty dual exhaust system, black cast wheels
FLSTN Heritage Softail Nostalgia - Limited edition based on FLST with two-tone Birch White & Black, small leather saddle bags
FLST/FXST/FLSTF - No major changes
FLHS/TC/U - Remote oil reservoir mounted under engine, battery under seat, 15% bigger saddlebags
FLTC/U - Remote oil reservoir mounted under engine, battery under seat, 15% bigger saddlebags

1994

The Range

94XLH 883 Sportster - 54ci (**883cc**) Evo V-twin, five-speed
94XLH 883 Sportster Hugger - 54ci Evo V-twin, five-speed
94XLH 883 Sportster Deluxe - 54ci Evo V-twin, five-speed
94XLH 1200 Sportster - 73ci (1203cc) Evo V-twin, five-speed

ABOVE: Some police departments stayed loyal to Harley-Davidson through the 1980s. This is a 1992 FXRP (denoting a police-spec Dyna Glide) in service in Daytona.

LEFT: Harley wasn't backward in celebrating its own long history, and this is a 90th-Anniversary FLH in 1993. Handlebar fairing and panniers were standard, but the top box was probably a factory extra. Note the generous passenger backrest.

OPPOSITE

LEFT: The FLHR Road King (this is the original 1994 version) was no more nor less than the latest Electra Glide, stripped of all the touring equipment apart from spotlights and panniers.

RIGHT: Some still use the big Harleys for camping trips, though the rider has left little room for a passenger. Twin front discs were standard by this time, but were universally criticized for not being effective in stopping a bike the weight of the Electra Glide.

94FLSTC Heritage Softail Classic- 82ci (**1340cc**) Evo V-twin, five-speed
94FLSTN Heritage Softail Special - 82ci Evo V-twin, five-speed
94FLSTF Fat Boy - 82ci Evo V-twin, five-speed
94FXR Super Glide - 82ci Evo V-twin, five-speed
94FXDS-CONV Low Rider Convertible - 82ci Evo V-twin, five-speed
94FXLR Low Rider Custom - 82ci Evo V-twin, five-speed
94FXSTC Softail Custom - 82ci Evo V-twin, five-speed
94FXSTS Springer Softail - 82ci Evo V-twin, five-speed
94FXDL Dyna Low Rider - 82ci Evo V-twin, five-speed
94FXDWG Dyna Wide Glide - 82ci Evo V-twin, five-speed
94FLHR Electra Glide Road King - 82ci Evo V-twin, five-speed
94FLHTC Electra Glide Classic - 82ci Evo V-twin, five-speed
94FLHTCU Electra Glide Classic Ultra - 82ci Evo V-twin, five-speed
94FLTC Tour Glide Classic - 82ci Evo V-twin, five-speed
94FLTCU Tour Glide Classic Ultra - 82ci Evo V-twin, five-speed

Total Production - 96,000

Colours

Vivid Black, Aqua Pearl, Victory Red Sun-Glo, Two-tone Aqua Pearl & Silver, two-tone Victory Red Sun-Glo & Platinum

Specifications

XLH883/1200 Sportsters - No major changes
FXR/LR/DWG/DL/STC/STS - Larger engine sprocket
FLSTN Heritage Softail Special - '93 FLSTN in two-tone Birch and Silver

FLSTF/STC - Larger engine sprocket
FLHR Electra Glide Road King - New model based on FLHTC, without fairing. Quickly-detachable windscreen and pillion seat, air suspension, large headlight in nacelle, twin front disc brakes, cast wheels
FLHTC/U - Larger engine sprocket, sealed cruise switch (U only)
FLTC/U - Larger engine sprocket, sealed cruise switch (U only)

1995

The Range

95XLH 883 Sportster - 54ci (**883cc**) Evo V-twin, five-speed
95XLH 883 Sportster Hugger - 54ci Evo V-twin, five-speed
95XLH 883 Sportster Deluxe - 54ci Evo V-twin, five-speed
95XLH 1200 Sportster - 73ci (1203cc) Evo V-twin, five-speed
95FLSTC Heritage Softail Classic - 82ci (**1340cc**) Evo V-twin, five-speed
95FLSTN Heritage Softail Special - 82ci Evo V-twin, five-speed
95FLSTF Fat Boy - 82ci Evo V-twin, five-speed
95FXD Dyna Super Glide - 82ci Evo V-twin, five-speed
95FXDS-CONV Low Rider Convertible - 82ci Evo V-twin, five-speed
95FXSTSB Bad Boy - 82ci Evo V-twin, five-speed
95FXSTC Softail Custom - 82ci Evo V-

twin, five-speed
95FXSTS Springer Softail - 82ci Evo V-twin, five-speed
95FXDL Dyna Low Rider - 82ci Evo V-twin, five-speed
95FXDWG Dyna Wide Glide - 82ci Evo V-twin, five-speed
95FLHT Electra Glide Standard - 82ci Evo V-twin, five-speed
95FLHR Electra Glide Road King - 82ci Evo V-twin, five-speed
95FLHTC Electra Glide Classic - 82ci Evo V-twin, five-speed
95FLHTCU Electra Glide Classic Ultra - 82ci Evo V-twin, five-speed
95FLHTCUI Electra Glide Classic Ultra - 82ci Evo V-twin fuel injection, five-speed
95FLTCU Tour Glide Classic Ultra - 82ci Evo V-twin, five-speed

Sample Prices

XL883 Sportster Deluxe - $6,120
FXD Dyna Super Glide - $9,995
FXSTB Bad Boy - $13,850
FLHTCU Electra Glide Classic Ultra - $16,050

Total Production - 105,000

Colours

Single tones: Vivid Black, Aqua Pearl, Victory Red Sun-Glo
Two-tones: Aqua Pearl & Silver, Nugget Yellow & Birch White, Charcoal Satinbrite & Vivid Black, Vivid Black & Silver, Birch White & Gold, Burgundy Pearl & Scarlet Red, Vivid Black &Burgundy Pearl

Specifications

XLH883/1200 Sportsters - Electronic speedometer
FXD Dyna Super Glide - New base-model Super Glide, using Dyna Glide chassis. Staggered shorty dual exhaust, cast wheels, electronic speedometer
FXDS-CONV/DWG/DL - Electronic speedometer
FXSTSB - New model based on FXSTS. Black steel springer forks, staggered shorty dual exhaust, spoked front wheel, slotted disc rear wheel, floating brake calipers
FLHT Electra Glide Standard - new base-model of FLHTC. Cast wheels, tachometer, fuel gauge, voltmeter, oil-pressure gauge, air suspension, fork-mounted fairing, lockable saddlebags, luggage rack, accessory plug
FLHTCUI - Sequential port fuel injection

1996

The Range

96XLH 883 Sportster - 54ci (**883cc**) Evo V-twin, five-speed
96XLH 883 Sportster Hugger - 54ci Evo V-twin, five-speed
96XLH 1200 Sportster - 73ci (**1203cc**) Evo V-twin, five-speed
96XL 1200C Sportster Custom - 73ci Evo V-twin, five-speed
96XL 1200S Sportster Sport - 73ci Evo V-twin, five-speed
96FLSTC Heritage Softail Classic- 82ci (**1340cc**) Evo V-twin, five-speed
96FLSTN Heritage Softail Special - 82ci Evo V-twin, five-speed
96FLSTF Fat Boy - 82ci Evo V-twin, five-speed
96FXD Dyna Super Glide - 82ci Evo V-twin, five-speed
96FXDS-CONV Low Rider Convertible - 82ci Evo V-twin, five-speed
96FXSTSB Bad Boy - 82ci Evo V-twin, five-speed
96FXSTC Softail Custom - 82ci Evo V-twin, five-speed
96FXSTS Springer Softail - 82ci Evo V-twin, five-speed
96FXDL Dyna Low Rider - 82ci Evo V-twin, five-speed
96FXDWG Dyna Wide Glide - 82ci Evo V-twin, five-speed
96FLHT Electra Glide Standard - 82ci Evo V-twin, five-speed
96FLHR Electra Glide Road King - 82ci Evo V-twin, five-speed
96FLHRI Electra Glide Road King - 82ci Evo V-twin fuel injection, five-speed
96FLHTC Electra Glide Classic - 82ci Evo V-twin, five-speed
96FLHTCI Electra Glide Classic - 82ci Evo V-twin fuel injection, five-speed
96FLHTCU Electra Glide Classic Ultra - 82ci Evo V-twin, five-speed
96FLHTCUI Electra Glide Classic Ultra - 82ci Evo V-twin fuel injection, five-speed
96FLTCUI Tour Glide Classic Ultra - 82ci Evo V-twin fuel injection, five-speed

Sample Prices

XL883 Sportster - $5,095
XL1200S Sportster Sport - $7,910
FXDL Dyna Low Rider - $13,030
FXDWG Wide Glide - $14,030
FLTCUI Tour Glide Classic Ultra - $17,410

Total Production - 119,000

Colours

Single tones: Vivid Black, Patriot Red Pearl, States Blue Pearl, Platinum Silver, Mystique Green Metallic, Violet Pearl
Two-tones: Victory Sun-glo & Platinum, Violet & Red Pearl, Platinum Silver & Black, Mystique Green &Black,
FXSTSB - Vivid Black with yellow, purple or turquoise graphics

Specifications

XLH883/1200 Sportsters - 3.3-gal (15-litre) fuel tank
XL1200S Sportster Sport - New model, based on XLH1200. Adjustable front forks, adjustable rear shocks, 3.3-gal tank
XL1200C Sportster Custom - New model, based on XLH1200. Staggered shorty dual exhaust, spoked front wheel, disc rear wheel
FXSTC/STS/B - Electronic speedometer
FLSTN/STC - Electronic speedometer

OPPOSITE: The 1996 FLHT Electra Glide Standard was a new 'entry-level' version of the big tourer, less expensive than the Classic and Ultra versions, but still with the handlebar fairing and luggage. However, it was a starting point for any serious tourist.

ABOVE: By 1997, fuel injection was an option on the Electra Glide, though this one is the carburettored version. The object of injection was not increased power, but cleaner emissions and more consistent running, especially from cold. The Electra Glide had long since ducked out of the power game, assuming it had ever been in it.

1997

The Range

97XLH 883 Sportster - 54ci (**883cc**) Evo V-twin, five-speed
97XLH 883 Sportster Hugger - 54ci Evo V-twin, five-speed
97XLH 1200 Sportster - 73ci (**1203cc**) Evo V-twin, five-speed
97XL 1200C Sportster Custom - 73ci Evo V-twin, five-speed
97XL 1200S Sportster Sport - 73ci Evo V-twin, five-speed
97FLSTC Heritage Softail Classic - 82ci (**1340cc**) Evo V-twin, five-speed
97FLSTS Heritage Springer - 82ci Evo V-twin, five-speed
97FLSTF Fat Boy - 82ci Evo V-twin, five-speed
97FXD Dyna Super Glide - 82ci Evo V-twin, five-speed
97FXDS-CONV Low Rider Convertible - 82ci Evo V-twin, five-speed
97FXSTSB Bad Boy - 82ci Evo V-twin, five-speed
97FXSTC Softail Custom - 82ci Evo V-twin, five-speed
97FXSTS Springer Softail - 82ci Evo V-twin, five-speed
97FXDL Dyna Low Rider - 82ci Evo V-twin, five-speed
97FXDWG Dyna Wide Glide - 82ci Evo V-twin, five-speed
97FLHT Electra Glide Standard - 82ci Evo V-twin, five-speed
97FLHR Electra Glide Road King - 82ci Evo V-twin, five-speed
97FLHRI Electra Glide Road King - 82ci Evo V-twin fuel injection, five-speed
97FLHTC Electra Glide Classic - 82ci Evo V-twin, five-speed
97FLHTCI Electra Glide Classic - 82ci Evo V-twin fuel injection, five-speed
97FLHTCU Electra Glide Classic Ultra - 82ci Evo V-twin, five-speed
97FLHTCUI Electra Glide Classic Ultra - 82ci Evo V-twin fuel injection, five-speed
97FLTCUI Tour Glide Classic Ultra - 82ci Evo V-twin fuel injection, five-speed
Total Production - 132,000

1998

The Range

98XLH 883 Sportster - 54ci (**883cc**) Evo V-twin, five-speed
98XLH 883 Sportster Hugger - 54ci Evo V-twin, five-speed
98XL53C Sportster Custom 53 - 54ci Evo V-twin, five-speed
98XLH 1200 Sportster - 73ci (**1203cc**) Evo V-twin, five-speed
98XL 1200C Sportster Custom - 73ci Evo V-twin, five-speed
98XL 1200S Sportster Sport - 73ci Evo V-twin, five-speed
98FLSTC Heritage Softail Classic - 82ci (**1340cc**) Evo V-twin, five-speed
98FLSTF Fat Boy - 82ci Evo V-twin, five-speed
98FLSTS Heritage Springer - 82ci Evo V-twin, five-speed
98FXD Dyna Super Glide - 82ci Evo V-twin, five-speed
98FXDS-CONV Low Rider Convertible - 82ci Evo V-twin, five-speed
98FXSTC Softail Custom - 82ci Evo V-twin, five-speed
98FXSTS Springer Softail - 82ci Evo V-twin, five-speed
98FXSTB Night Train - 82ci Evo V-twin, five-speed
98FXDL Dyna Low Rider - 82ci Evo V-twin, five-speed
98FXDWG Dyna Wide Glide - 82ci Evo V-twin, five-speed
98FLHT Electra Glide Standard - 82ci Evo V-twin, five-speed
98FLHR Road King - 82ci Evo V-twin, five-speed
98FLHRCI Road King Classic - 82ci Evo V-twin fuel injection, five-speed
98FLHP Road King Police - 82ci Evo V-twin, five-speed
98FLHPI Road King Police EFI - 82ci Evo V-twin, fuel injection, five-speed
98FLTR Road Glide - 82ci Evo V-twin, five-speed
98FLTRI Road Glide EFI - 82ci Evo V-twin fuel injection, five-speed
98FLHTC Electra Glide Classic - 82ci Evo V-twin, five-speed
98FLHTCI Electra Glide Classic - 82ci Evo V-twin fuel injection, five-speed
98FLHTCUI Electra Glide Classic Ultra - 82ci Evo V-twin fuel injection, five-speed
98FLHTPI Electra Glide Police EFI - 82ci Evo V-twin fuel injection, five-sped
Total Production - 150,000

Colours

Single tones: Vivid Black, Lazer Red Pearl, Sinister Blue Pearl, Violet Pearl
Two-tones: Lazer Red & Black Midnight, Red & Champagne Pearl, Sinister Blue

Pearl & Platinum Silver, Mystique Green & Black

Specifications

XLH883/1200 Sportsters - New rear light, new oil pump (increased capacity), electronic speedometer (XLH883 only), integrated timer sensor/ignition module (except 1200S), 1200S gets 15% torque increase from 4-pole coil, 2-plug heads, 10:1 compression, less restrictive exhaust, new camshaft, new ignition
XL53C Sportster Custom - New model, based on basic 883 with chrome details, bullet headlight, disc 16-inch (406-mm) rear wheels and spoked 21-inch (533-mm) front wheel, twin seat
FXD/DL/CONV/WG - Nine-plate clutch, new tail light and neutral switch
FXSTC/STS/STF/FLSTS - Nine-plate clutch, neutral switch
FXSTB Night Train - New model, based on FXSTC, Europe-only for the first year. All-black Evo V-twin, black theme for oil tank, belt guard, mirrors etc, drag bars, forward control, minimal seat, wide front forks, 21-inch front wheel, 16-inch disc back rear wheel
FLHCRI Road King Classic - New model, based on FLHR, with leather saddlebags and slant-cut silencers
FLTRI Road Glide - New model to replace Tour Glide, comes in stripped-down form with no Tour-Pak, saddlebag guard rails or fender bumpers, new fairing, low windscreen, new seat, cast wheels, new sound system, gauges for fuel, ambient temp, volt, oil pressure
All FLs - Nine-plate clutch, upgraded cruise control, neutral switch, new sound system
95th Anniversary models - XL1200C, FXDWG, FLSTF, FLSTS, FLHTCI, TLHTCUI, FLHRCI, FLTRI - special graphics, individually numbered plates, Two-tone Midnight Red/Champagne Pearl

1999

The Range

99XLH 883 Sportster - 54ci (**883cc**) Evo V-twin, five-speed
99XLH 883 Sportster Hugger - 54ci Evo V-twin, five-speed
99XL53C Sportster Custom 53 - 54ci Evo V-twin, five-speed
99XLH 1200 Sportster - 73ci (**1203cc**) Evo V-twin, five-speed
99XL 1200C Sportster Custom - 73ciEvo V-twin, five-speed
99XL 1200S Sportster Sport - 73ci Evo V-twin, five-speed
99FLSTC Heritage Softail Classic - 82ci (**1340cc**) Evo V-twin, five-speed
99FLSTF Fat Boy - 82ci Evo V-twin, five-speed
99FLSTS Heritage Springer - 82ci Evo V-twin, five-speed
99FXD Dyna Super Glide - 88ci (**1450cc**) Twin Cam V-twin, five-speed
99FXDX Dyna Super Glide Sport - 88ci Twin Cam V-twin, five-speed
99FXST Softail Standard - 88ci Twin Cam V-twin, five-speed
99FXSTB Night Train - 82ci Evo V-twin, five-speed
99FXSTS Springer Softail - 82ci Evo V-twin, five-speed
99FXDL Dyna Low Rider - 88ci Twin Cam V-twin, five-speed
99FXDWG Dyna Wide Glide - 88ci Twin Cam V-twin, five-speed
99FLHT Electra Glide Standard - 88ci Twin Cam V-twin, five-speed
99FLHR Road King - 88ci Twin Cam V-twin, five-speed

99FLHRCI Road King Classic - 88ci Twin Cam V-twin fuel injection, five-speed
99FLTRI Road Glide EFI - 88ci Twin Cam V-twin fuel injection, five-speed
Total Production - 170,000
Colours
Single tones: Vivid Black, Lazer Red Pearl, Aztec Orange Pearl, Cobalt Blue Pearl, Diamond Ice Pearl, Nugget Yellow Pearl, Scarlet Red Pearl, White Pearl
Two-tones: Sinister Blue & Diamond Ice, Aztec Orange & Diamond Ice, Mystic Green & Black, Lazer Red & Black
Specifications
XLH883/1200 Sportsters - No major changes
XL53C - Longer front forks, drag bars and forward controls
XL1200C - Drag bars, forward controls
FXD/DL/WG - Twin-Cam 88 V-twin, 3.74 x 4in (95 x 102mm), 88ci, carburettor, rubber-mounted, claimed more power/torque, better reliability, stiffer engine/gearbox joint, redesigned lubrication system, quieter
FXDX - New model, based on FXD. Dyna chassis with Twin Cam 88 V-twin in all-black form, black XR750-style handlebars, twin front discs, spoked wheels, raked forks
FXSTS/STB/FLSTF/STS/FTC - New speedometer to allow on-board engine diagnostics
FLHR/CI/CUI/FLTRI - Twin Cam 88 V-twin for all the tourers, integrated rear brake reservoir and master cylinder, speedometer to allow on-board engine diagnostics (FLHR/CI only)

2000

The Range
00XL883 Sportster - 54ci (**883cc**) Evo V-twin, five-speed
00XL883 Sportster Hugger - 54ci Evo V-twin, five-speed
00XL53C Sportster Custom 53 - 54ci Evo V-twin, five-speed
00XL1200C Sportster Custom - 73ci (**1203cc**) Evo V-twin, five-speed
00XL1200S Sportster Sport - 73ci Evo V-twin, five-speed
00FLSTC Heritage Softail Classic - 88ci (**1450cc**) Twin Cam 88B V-twin, five-speed
00FLSTF Fat Boy - 88ci Twin Cam 88B V-twin, five-speed
00FXD Dyna Super Glide - 88ci Twin Cam V-twin, five-speed
00FXDX Dyna Super Glide Sport - 88ci Twin Cam V-twin, five-speed
00FXDL Dyna Low Rider - 88ci Twin Cam V-twin, five-speed
00FXDWG Dyna Wide Glide - 88ci Twin Cam V-twin, five-speed
00FXST Softail Standard - 88ci Twin Cam 88B V-twin, five-speed
00FXSTB Night Train - 88ci Twin Cam 88B V-twin, five-speed
00FXSTD Softail Deuce - 88ci Twin Cam V-twin, five-speed
00FLHR Road King - 88ci Twin Cam V-twin, five-speed
00FLTRI Road Glide - 88ci Twin Cam V-twin fuel injection, five-speed
00FLHRCI Road King Classic - 88ci Twin Cam V-twin fuel injection, five-speed
00FLHTCUI Electra Glide Classic Ultra - 88ci Twin Cam V-twin fuel injection, five-speed
Total Production - 203,000 (projected from Sept. 2000)
Colours
Single tones: Vivid Black, Chrome Yellow, Aztec Orange Pearl, Cobalt Blue Pearl, Luxury Rich Red, Bronze Pearl, Diamond Ice Pearl, Sinister Blue, Concord Purple, Suede Green Pearl, White Pearl,
Two-tones: Luxury Rich Red/Vivid Black, Sinister Blue/Diamond Ice, Aztec Orange/Diamond Ice, Vivid Black/Chrome Yellow, Suede Green/ Vivid Black
Specifications
XLH883/1200 Sportsters - Four-pot front disc brake, sealed wheel bearings, sealed battery, stiffer crankshaft, redesigned tappet blocks, new transmission internals (smoother gearchange). XL53C and 1200C have 3.7-gal (16.8-litre) Fat Bob fuel tank
FXD/DL/WG - Four-pot front disc brake, sealed wheel bearings, sealed battery, new drive belt, new transmission internals, redesigned belt guard, swinging arm, fuse system
FXDX - As above, plus cartridge-type front forks with adjustable pre-load, rebound and compression damping, gas-charged rear shocks with dual rate springs and adjustment for pre-load and rebound damping, wider tyres, redesigned seat, cast alloy wheel option, new sidestand
All Softails - New Twin Cam 88B V-twin, with twin counter-rotating balancer shafts to reduce vibration
FXST/B/D/FLSTF/STC - Four-pot front disc brake, sealed wheel bearings, sealed battery

OPPOSITE:
LEFT: The 1999 Electra Glide Ultra Classic, complete with its new Twin Cam V-twin. Despite the name, the latest Harley twin does not have twin overhead cams, though it is a clear advance on the Evo. A balance shaft version is soon to follow.

RIGHT: King of the Retros: the Springer Softail (this is a 1988 version) really was the ultimate nostalgia machine, combining the Softail rear end with an updated version of Bill Harley's original 1905 springer front fork. It didn't work as well as a modern telescopic fork, but does that really matter?

BELOW: The five-speed 2001 FXSTS Springer Softail with 88ci (1450-cc) Twin Cam 88B.

FXSTD - New model, based on FXST. Stretched 4-gal (18-litre) tank, chrome front end, disc rear wheel, new rear mudguard
FLSTF - Repositioned gear lever, redesigned exhaust (easier removal) and rear mudguard
FLHR/CI/TCUI/FLTRI - Four-pot front disc brake, sealed wheel bearings, sealed battery, new drive belt, new transmission internals, 9-spoke cast alloy wheels, new frame (easier access to engine top end)

2001

The Range

01XL883 Sportster - 54ci (**883cc**) Evo V-twin, five-speed
01XL883 Sportster Hugger - 54ci Evo V-twin, five-speed
01XL53C Sportster Custom 53 - 54ciEvo V-twin, five-speed
01XL1200C Sportster Custom - 73ci (**1203cc**) Evo V-twin, five-speed
01XL1200S Sportster Sport - 73ci Evo V-twin, five-speed
01FLSTC Heritage Softail Classic - 88ci (**1440cc**) Twin Cam 88B V-twin, five-speed
01FLSTCI Heritage Softail Classic EFI - 88ci Twin Cam 88B V-twin fuel injection, five-speed
01FLSTS Heritage Springer - 88ci Twin Cam 88B V-twin, five-speed
01FLSTF Fat Boy - 88ci Twin Cam 88B V-twin, five-speed
01FLSTFI Fat Boy EFI - 88ci Twin Cam 88B V-twin fuel injection, five-speed
01FXD Dyna Super Glide - 88ci Twin Cam V-twin, five-speed
01FXDX Dyna Super Glide Sport - 88ci Twin Cam V-twin, five-speed
01FXDXT Dyna Super Glide T-Sport - 1,449cc Twin Cam V-twin, five-speed
01FXDL Dyna Low Rider - 88ci Twin Cam V-twin, five-speed
01FXDWG Dyna Wide Glide - 88ci Twin Cam V-twin, five-speed
01FXST Softail Standard - 88ci Twin Cam 88B V-twin, five-speed
01FXSTB Night Train - 88ci Twin Cam V-twin 88B, five-speed
01FXSTD Softail Deuce - 88ci Twin Cam 88B V-twin, five-speed
01FXSTDI Softail Deuce EFI - 88ci Twin Cam 88B V-twin fuel injection, five-speed
01FXSTS Springer Softail - 88ci Twin Cam 88B, five-speed
01FLHR Road King - 88ci Twin Cam V-twin, five-speed
01FLTRI Road Glide - 88ci Twin Cam V-twin fuel injection, five-speed
01FLHRCI Road King Classic - 88ci Twin Cam V-twin fuel injection, five-speed
01FLHT Electra Glide Standard - 88ci Twin Cam V-twin, five-speed
01FLHTCUI Electra Glide Classic Ultra - 88ci Twin Cam V-twin fuel injection, five-speed

Colours

Single tones: Vivid Black, Luxury Blue Pearl, Jade Sunglo Pearl, White Pearl, Luxury Rich Red Pearl, Chrome Yellow Pearl, Real Teal Pearl, Concord Purple

LEFT: 2001 Springer Softail, with an updated version of Bill Harley's original 1905 springer front fork.

ABOVE: Close-up of a 2008 Dyna Street Bob. Picture courtesy of Flickr/Danie VDM.

OPPOSITE: A 2008 Classic Electra Glide. Picture courtesy of WorldWideMotorcycles.

Pearl, Bronze Pearl, Diamond Ice Pearl, Suede Green Pearl
Two-tones: Luxury Blue & Diamond Ice, Luxury Rich Red & Black, Vivid Black & Chrome Yellow, Concord Purple & Diamond Ice, Real Teal & Birch White, Vivid Black & Chrome Yellow, Suede Green & Black

Specifications

All Sportsters - New oil pump, Spinylok cylinder liners to reduce oil consumption
XL883 Sportsters - 7% power increase from revised induction/exhaust systems
All Dyna Glides - New fuse panel, easier to use side stand, improved gearbox internal shifters, factory fitted security/immobilizer system, Dunlop H-D tyres
FXDL Low Rider - Lower seat (from shorter rear shocks), low-rise bars on pull-back risers
FXDWG Wide Glide - New seat and backrest
FXDXT T-Sport - New model, using Twin Cam 88B balance-shaft V-twin, twin front discs, adjustable front and rear suspension, fork-mounted fairing, adjustable screen, ballistic nylon saddlebags
All Softails - New engine diagnostics, factory fitted security/immobiliser systems, EFI option on Fat Boy, Deuce and Heritage Softail Classic
FLH/T - Improved gearbox internal shifters, factory fitted security/immobilizer system, new engine diagnostics, Dunlop H-D tyres, metal air dam below steering head removed to improve hot weather comfort

2002-2010

Leading on to the end of the decade, the Harley-Davidson brand is as strong as ever. Manufacturing has continued with the Sportster, Dyna, Softail, VRSC and Touring models.

HARLEY-DAVIDSON LIFESTYLE

What then is the work of Life? What the business of great men, that pass the stage of the world in seeming triumph as these men, we call heroes, have done? Is it to grow great in the mouth of fame and take up many pages in history? Or is their business rather to add virtue and piety to their glory, which alone will pass them into eternity and make them truly immortal? Thus wrote Daniel Defoe, 'On the Instability of Human Glory'.

It is unlikely when William S. Harley and Arthur Davidson of Milwaukee, Wisconsin produced their first motorcycle in 1903, that they had any notion of how famous they would eventually become, particularly as there is a debate as to their initial intentions. Did they produce the bike entirely for themselves, or was it to be the first of many, and the start of a serious business venture?

Whatever the truth, even when they began to receive requests for bikes, they would have had no idea that they were founding a dynasty and creating a legend that would become a 21st-century icon.

That first Harley-Davidson, like those that were to follow, was built to be strong and reliable on roads that were far from

LEFT: HOG riders in South Africa. Picture courtesy of Flickr/DanieVDM

OPPOSITE: Sturgis: riding out to Wyoming and the Devil's Tower.

deal for rudimentary forms of motorized transport. However, that initial policy of strength and reliability was to serve them in good stead and enable Harley and the Davidsons to stay in business as other firms appeared and then went to the wall.

In 1909 when the first V-twin was added, the beginning of the legend was consolidated. 'The Motor Company', as it was known in-house, has been faithful to the same basic format ever since, through thick and thin, good times and bad, because it has stuck to its initial principles. As a result the motorcyclists, Harley's buying public, have remained faithful in their turn.

In this way, a bond has been created and a fierce brand loyalty has developed that has spilled out onto the social scene and society in general, creating a unique

lifestyle to evolve around the name Harley-Davidson, which is probably one of the best known the world over. People have it tattooed across their bodies; they hang posters on their walls; and apply the emblem wherever there is space. They even buy books on the subject!

And this doesn't only apply to motorcyclists. There are those with no interest whatsoever in bikes with whom the name Harley-Davidson will strike an immediate chord and, if asked to name a bike, is the first that would spring to mind. Watch any disinterested bystander when a Harley booms into view: they turn, they look and stand, mouth open, as the bike heads out of focus into the far horizon. This does not happen with a Honda step-through.

Harley-Davidson has come to stand for the American way of life as it is lived on two wheels. It may only be a bike, but to most it offers a sense of freedom that many feel is missing from their busy lives. Maybe it is this factor more than anything else that has led to the advent of the weekend bikers, or 'rebels with a cause'.

On first sight, fake henna tattoos and stick-on ponytails may make them look different, but a second glance will confirm that they are also wearing smart tailored leathers: they do not want their true status and affluence to go unnoticed. The Harley has become a style icon for the 'rubbies', RUBs or rich urban bikers, to give them their full title. They may drive an executive car from Monday to Friday, but once the weekend arrives, a top-of-the-range Harley will suddenly appear and the transformation is complete, from city slicker to weekend rebel at the lift of a garage door.

However, at the other end of the scale is what many would call the true biker.

LEFT: The Harley Owners Group (HOG) is their officially sanctioned club, with chapters existing all over the world.

OPPOSITE: Do you feel brave enough to ask this man what he thinks of the latest Honda FireBlade?

These are associated with the patch clubs and would have you believe they are one of the 'one percenters'. For these riders, their Harley is not a leisure tool to be kept under covers during the week; it is an important part of their life, something which defines them and connects them to their associates. This is the motorcycle lifestyle which has been fully embraced.

Both of these groups are at the extreme limits of the Harley lifestyle, one so diverse that it encompasses every facet of society and its products. There are goods, clothing and other items in such abundance that one wonders what their connection is with the world's most famous motorcycle, apart from earning the royalties that tumble into the corporate coffers.

While it is an easy matter to trace the origins of the Harley-Davidson and to chart its progress by means of particular models, race successes, or company decisions, what is far more difficult is to understand how the Harley lifestyle phenomon came into being.

Opinions differ as to what the initial impetus was, but there is one factor which has helped consolidate the Harley image: namely the media. Perhaps the best illustration of this is the infamous Hollister incident of 1947, which many claim to be

HARLEY-DAVIDSON
BIKE WEEK

the beginning of the back-patch gangs and anti-motorcycle propaganda, and which led to one of the film industry's most famous productions, *The Wild One*, starring Marlon Brando.

The truth is that the events at Hollister had their roots in the Second World War, or at least its ending. During the war, many US servicemen learned to ride on a Harley WLA. Freshly demobbed in the mid-1940s, many suddenly found themselves abruptly returned to 'normal' life, and deprived of the comradeship and element of danger that had become a way of life. Those unrestricted by jobs or families naturally drifted, and becoming a bike gang member was something that echoed many of their wartime experiences: the kudos associated with being something of an outsider, and being surrounded with people with whom you could completely identify.

Meanwhile, the American Motorcycle Association (AMA) had decided to resurrect its Gypsy Tour, a race series current throughout the US that took in flat tracks and hill climbs that had been popular prior to the war. With cheap gasoline and nothing much to tie them to any one place, these groups of riders headed for the races to watch and compete. There was little in the way of hotels, the motel concept having not yet caught on, so accommodation and public facilities in general were sparse apart from the big cities. Riders would often camp out or sleep rough in small towns that simply couldn't cope with the influx.

So it was with Hollister, California, a town of just 4,000 inhabitants at the time, whose only claim to fame was that it was America's biggest producer of garlic. On the Fourth of July weekend in 1947 they were playing host to an AMA-sanctioned event that was part of the new Gypsy Tour.

ABOVE: A rally enthusiast: every badge tells a story.

RIGHT: Pets are allowed at Sturgis, so now you can see the point of those broad Fat Bob tanks.

OPPOSITE: Flawless tans and big hair feature prominently at Sturgis.

MOTOR
HARLEY-DAVIDSON
CYCLES

It was to be held on the banked $^{1}/_{3}$-mile oval on the outskirts of town.

On the Friday night, the town began to fill with riders, many arriving to compete in the hill climb that was to begin the meeting. By Saturday, the population had doubled in size for the main races, which saw Harley riders, Joe Leonard in particular, dominating the proceedings.

But while all was going well with the racing, matters were deteriorating down in town. Many refused to attend the official function in the evening, but instead virtually took over San Benito Street, where most of the town's bars and restaurants were situated. The atmosphere became saturated with carbon monoxide as unofficial drag races were staged between gang members, some of which were doing the rounds up and down the street. They had colourful names like the Boozefighters, Yellow Jackets, Satan's Sinners, 13 Rebels, Winos and so on. Funnily enough, not all of these were riding Harleys.

The local police began to feel they were losing control and called in the California Highway Patrol. They had had the sense to realize that they would be unable to deal with the situation singlehanded and, showing considerable imagination, hired a loud band to play on the outskirts of town, the idea being to lure people away from the town centre. This did not go quite according to plan, and scuffles ensued which were later described as pitched battles. The result was that 50 people were arrested for offences ranging from urinating in the street to drunkenness, indecent exposure and resisting arrest. Unpleasant though all this may sound, however, it hardly amounted to murder.

But in a small town like Hollister, it was a sensation. Whether or not the police had been heavy-handed hardly mattered.

OPPOSITE and ABOVE: Sturgis police have a choice of mounts when it comes to cruising the shows.

RIGHT: Crossing the flatlands on a Heritage Softail.

A memorial day party, the Red River Run, Mexico.

Cruising Daytona: it doesn't have to be along the beach.

Something big was going on, and when the editor of the *San Francisco Chronicle* got wind of it a reporter and photographer were sent out by chartered plane. (Some say he was tipped off by a police radio dispatcher, while according to others, they arrived merely by chance!)

However, by the time they arrived it was all over, but having put in so much effort to get there, they persuaded one of the revellers to pose, beer bottle in hand, astride his Harley outside Johnnie's bar, which had been the centre of the action. It was to become one of the most famous shots ever when it appeared all over the US, in newspapers, magazines and eventually in books. Following the report and picture appearing in the *Chronicle* it was also picked up by *Life* magazine who used it as a front cover shot. All this served to exaggerate the events at Hollister which began by being described as a riot, and ended with a quote from the police chief saying 'It's just one hell of a mess', which somewhat strengthened the story.

Within a week, everyone riding a motorcycle was typecast a fat overweight, bleary-eyed 'Oakland Boozefighter', out to capture every good-looking women

LEFT: Many people still ride to Daytona, stopping for fuel and refreshment along the road.

OPPOSITE
LEFT: Cruising on a hand-change, foot-clutch Harley in Sundance, Wyoming.

RIGHT: A Harley was the first prize at this casino in Deadwood.

they encountered, mounted, of course, on Harley Davidsons: so the biker image was born and was associated with Harley ever after.

This was the true incident that Stanley Kubrick's film *The Wild One* further immortalized in 1953, a young Marlon Brando playing an introspective Johnny Strabler, the leader of an outlaw gang.

In fact, Harleys did not feature prominently in the film (Brando was riding a Triumph), which compared motorcyclists to cowboys. The unfortunate effect was to make the words biker and outlaw synonymous, with Harleys as instruments of terror.

Moreover, this image continued after Hollister, with numerous other reported gatherings of bikers all being tarred with the same brush. The fact that the word 'outlaw' had been applied to them caused the AMA secretary at the time, Lin A. Kuchler to utter a much quoted comment: 'The disreputable cyclists were possibly one percent of the total number of motorcyclists, only one percent are hoodlums and troublemakers.'

This statement, despite it being made with the best of intentions to support the average rider, caused resentment among many. As a result, the 1% symbol came to be worn as a badge of honour and gangs like the Hell's Angels, Road Rats and Satan's Slaves united under this banner.

The AMA continued with its damage-limitation policy, but by now it was too late, and the anti-motorcycle lobby was in full cry, while some of these 'one percenters' continued to fuel the fire with anti-social behaviour and reckless riding. Meanwhile, Harley-Davidson attempted to distance itself from the trouble by issuing statements, but privately the company knew that this somewhat negative interest in motorcycles was doing it no harm at all,

LEFT: Ready to go!

ABOVE: There are no age limits when it comes to riding.

OPPOSITE: Hunlett, Wyoming, where the queue for the Ponderosa outstrips the queue of bikes.

LIQUOR
PONDEROSA
cafe BAR
Coors

and was helping it to sell bikes at a time when business was slack.

Japanese manufacturer Honda eventually spent a great deal of money on an advertising campaign claiming that 'You meet the nicest people on a Honda'. This did something towards changing public opinion and softened the motorcycling image to a certain degree. It also caused Harley to reassess its position. As the 1960s progressed, a new breed of rider was appearing who liked the style, but required a clean image rather than a criminal record.

Now to be seen outside the smartest cafés and shopping in town for designer clothes, motorcycle pilots began to wonder why it was necessary to take long swinging kicks to get a magneto to breathe life into a cold iron engine. Instead they required a quick press of a button to avoid breaking into a sweat or creasing their designer jeans. They wanted to be able to 'head out on the highway looking for adventure', as Steppenwolf would have said, but stay in smart hotels and motels along the way. This was the start of the weekend biker, and from this time forward Harley had, not only to market a bike, but an image as well. What follows is just a small look at what that involved and how the bikes came to appear along with all the other associated items and events that make up the 'Harley Lifestyle'.

Cruising is an integral part of big events like Sturgis and Daytona. Of course, people go there to meet old friends, make new ones, see the displays and watch the racing. But for many, the main reason for going is the cruise, not only to pose, but also to watch others do the same thing. It's the ultimate in audience participation, for anyone can climb on their bike and be an instant part of the spectacle.

LEFT: Just to prove you were there!
ABOVE: This is what they came to see: Devil's Tower, Wyoming.

LEFT: An Angel rides his Harley chop through an English autumn landscape.

BELOW: Stay cool and look mean.

OPPOSITE
LEFT: Sonny Barger with his own FXRS.

RIGHT: Don't be fooled, by Monday morning he'll be back in suit and tie.

SONNY BARGER'S

The Movie Industry

Like motorcycles and automobiles, the film industry at the turn of the 20th century was in its infancy and like the car industry, which tended to base itself in one city, so Hollywood became the home of the movies. While many different companies produced various films, there was understandably quite a large cross-pollenation of ideas.

The prop and set suppliers established their businesses in the area and consequently the same props appeared in many of the films. As any avid movie buff will confirm, every American town or city had a large name sign outside it and seated behind this sign was inevitably a police motorcyclist mounted on a Harley-Davidson, ready to pursue the hero or the bad guy, depending on the story line. While this may be artistic licence, it was based on fact: Harley sales to the police were good, so there was an element of truth with which the viewer could identify.

There was another reason why bikes always managed to feature so strongly; the movie moguls and stars were often keen riders and nearly always went for Harleys, Clark Gable being such a one long before he became a movie star.

In the 1930 film, *Hell's Angels*, made by Howard Hughes, the heroes are seen riding into town and back again on Harleys before embarking on a suicide mission to Germany. A prop maybe, but historically correct, for a large numbers of pilots also rode bikes, making the association between the two machines in their own minds. Three years on, a Harley Model J, complete with a sidecar, appeared in the Marx brothers' film, *Duck Soup*, while another outfit was used in the the Three Stooges comedy, *Playing the Ponies* in 1937.

But it was not until after the war that the rash of films with bikes appearing as the centre of attention really began to take hold. We have already mentioned *The Wild One*, but the success of the Kubrick movie, and the rise of the motorcycle bad-guy image that resulted from Hollister, encouraged an endless stream of biker movies, many of them coming well before *Easy Rider*: there were *Born Losers* 1967, *Hell's Angels on Wheels* 1967, *Angels from Hell* 1968, *Naked Under Leather* 1968, *The Glory Stompers* 1968, *The Angry Breed* 1969, and *Cycle Savages* 1969, to name but a few of the B-rated movies. There are actually over 70 of them listed in the film encyclopedia as having been made between 1960 and 1975, and in the vast majority a Harley-Davidson takes pride of place.

'Captain America', the bike ridden by Peter Fonda in *Easy Rider*, was the grandaddy of them all, with its stars-and-stripes paint job, raked out forks and Panhead motor. There were actually two of these bikes used in the film, one for long shots and another for close-ups. The latter was stolen and the other crashed in the final sequences, but countless Captain America replicas have since been built, for museums, shows and private individuals. The film itself was cut by the censor's scissors from 240 minutes to 94 and was a real budget movie, which Dennis Hopper made with $300,000. Despite this, it was and still is a cult film for many reasons and it helped reinforce the Harley myth. Just as *The Wild One* encouraged inferior follow-ons, so did *Easy Rider*, with *Angel Unchained* 1970, *Black Angels* 1970, *Chrome and Hot Leather* 1971, which were destined to slide into obscurity.

Electra Glide in Blue, a police-orientated film was different, and showed the bike in a more socially acceptable light. In the 1980s matters progressed even more with the film *Mask*, starring Cher. It was based on the true story of a woman and her disfigured son who have a dream to ride a Harley across Europe. However, and despite receiving much critical acclaim, it too slipped from main cinema viewing lists.

Since then, Harleys have continued to figure in action scenes. *Terminator 2 – Judgement Day*, with Arnold Schwarzenegger dwarfing the bike, springs immediately to mind. Then there was *Pulp Fiction*, *Made of Steel* and *Chrome Soldiers*: even Walt Disney used a Harley JD-H in *The Rocketeer*, showing that the image has changed just a little. Today though, Harley-Davidson has no problem with product placement. The bikes appear in films and TV shows primarily because of the powerful images the industry has created for them – one it is happy to continue to promote.

OPPOSITE: On the beach, Daytona.

BELOW: Gettng read for the off outside the Boothill Saloon, Daytona.

Events

There are a number of events that, while not always run purely for Harley aficionados, have become synonymous with the brand and the lifestyle image. Innumerable gatherings are organized by HOG (Harley Owners Group) and other Harley clubs. Here are just a few events that have become famous and are now internationally known.

Daytona Bike Week

On the east coast of Florida, nature has created a stretch of white sandy beach flanked by the Atlantic Ocean that makes it ideal for most of the year as a holiday resort for Americans – its name? Daytona. The actual city is split by an intercoastal waterway, leaving a long strip of land next to the beach that neatly separates this area from the main centres of commerce and industry.

In the early days of motoring, automobile races were held on that long white beach, but when cars got too fast, they departed for wider open spaces such as the Bonneville Salt Flats.

This hit Daytona hard, as the town was grateful for tourist income in the dead months of February and March, so it came up with a motorcycle race to fill the void – the 'Handlebar Derby'.

The first pure motorcycle race over

It doesn't all happen at the Boothill: this is the Iron Horse Saloon at Daytona.

You can almost smell the surf! The gathering of the faithful on Daytona beach.

MANGIA
PIZZA
248-2510
SICILIAN PIZZA
ITALIAN SUBS
Outlaws
mc
Florida

200 miles (322km) took place in 1937, with 120 riders racing on the part-road, part-beach track of 3.2miles (5km). While the field was liberally laced with Harleys, it was actually one of their rivals, an Indian, that took the honours in that inaugural meeting. It was such a success, that it became an annual event, shelved during the war, then restarted in 1947. The only problem was that races could only be run according to the vagaries of the tide, so in 1960 the last beach race was run (won by Brad Andres on a Harley) and the event moved 10 miles (16km) inland to a purpose-built stadium covering 450 acres (180 hectares) on the outskirts of the city. The first race here was also won on a

OPPOSITE: One percenters from Florida, at Daytona.

LEFT and ABOVE: Conflicting images at Daytona: sexist tattos can exist alongside women riding their own bikes and being anything but submissive. Above picture courtesy of Flickr/DeusXFlorida.

Harley ridden by Roger Reiman, whose name was to become synonymous with the marque.

Times were changing, however, and Harley was only ever to take the chequered flag another five times before the Japanese brands began to dominate, not only the classic 200-mile race, but also all the other races at Daytona.

But the coming of the Japanese was far from being the end of Harley-Davidson influence at Daytona – quite the reverse. While the racing had migrated inland, 'Bike Week', a massive seven-day meeting of like-minded bikers, remained on the narrow strip of land adjacent to the beach. It became an independent motorcycle scene, grew up, became one big party for bikers in general, and centred itself on Main Street, right in the centre of the spit. It hardly mattered that the beach racing had stopped: riders from all over the US would make it their home for a week, even while much of the rest of the US was covered in snow and ice. The vast majority never even set foot in either the main banked tri-oval circuit or, for that matter, the Municipal Stadium further down the road to watch the short-track racing.

Instead this was a chance for many to ride the 12-mile (19-km) stretch of the beach, hang out on Main Street, sink a few beers and just chill out. In 1973 the embryo 'Rat's Hole' custom show started and a year later 15,000 turned up to see the bikes on display, at a time when the police chief was expecting a third of that number.

This was also the first year that any trouble was recorded after the publication of the new motorcycle regulations for the city of Daytona. When these somewhat draconian laws were enforced, a minor backlash occurred resulting in a number of arrests, which the police were quick to play down.

Meanwhile, manufacturers like Harley-Davidson were realizing that Daytona was becoming much bigger than a mere race week and looked to setting up shop. Along Main Street, customizers and accessory suppliers took up short-term leases and shops and the crowds rolled in big time. In 1976, Harley took over a hotel for its exhibition and what with Harley drag races just down the road, the city was teeming with evidence of the marque: Daytona had become a second home for Harley-Davidson. And it had become an international event, with Harley riders arriving from all over the world. But matters were getting somewhat out of hand: the police turned a blind eye to the odd spot of impromptu beach racing, but in 1977 gang warfare broke out and a few shootings took place.

The following year all was quiet, but from then onward everything began to slide downhill. There were shoot-outs in bars and the one percenter element was rife. Gradually the police adopted a zero tolerance policy as visitors were encouraged to stay out of town by an element intent on breaking the week's annual monopoly of the American motorcycle scene. Many riders already felt they were being ripped off by hotels and traders in Main Street and needed little encouragement to stay away or move to the outlying towns. Custom bikes had to be hidden as they contravened the local laws, forcing the custom shows underground, and the whole Bike Week phenomenon was heading for collapse.

This occurred when a biker failed to stop for a police road block and was shot and killed. In court it emerged that the police were being trained and commanded to 'shoot to kill' bikers if there were any signs of misbehaviour, and the revelation resulted in the police chief and district attorney having to resign.

The new police chief adopted a different approach: he met the locals, worked out a plan of action and Bike Week returned to its former life as a party venue, but one which was now under control. However, much of its spontaneity was lost: it became too organized, even orchestrated, and was certainly over-commercialized.

But all this has not deterred over half a million flocking to Daytona every year for some winter sun.While there are now a few other makes appearing in the massive line- up and parades of bikes (even some Japanese sports bikes), Daytona is still very much Harley-dominated during Bike Week.

Standard, customized, chopped, it matters little, the parking lots and Main

OPPOSITE: Out for a quiet Sunday ride: a procession of riders at Daytona.

ABOVE: The classic Daytona image, and a classic bike to go with it.

Street reverberate to the thunder of tons and tons of Harley metal glistening in the sun as it rumbles by.

Policing is now a low-key affair that manages to keep a lid on things without upsetting the visitors. Arrests and citations are few and the main problem now revolves around traffic flow and the increasing number of accidents. This is inevitable considering the vast number of bikes that now criss-cross the city and run up and down the state visiting the various other attractions.

While racing may have been the start of the Daytona happening, it is now just a sideshow along with the other events that go to make up the week. For Harley-Davidson, meanwhile, Daytona is still a big event in the calendar, and it has taken over the massive Ocean centre for putting on displays, where test rides are run, and the end parade finishes at Harley Heaven. This is on the last Sunday of the week and signals the end of the partying before the town is inundated by the 'spring breakers' (vacationing students), who the police now regard as more troublesome than the motorcyclists. How times change!

RIGHT: A night scene: Harley-Davidson doesn't exactly dominate Daytona over Bike Week, but it comes very close to it.

OPPOSITE LEFT and RIGHT: Sturgis, where Main Street serves the same purpose as Daytona beach. It's the place to see and where one can be seen. But ride a couple of blocks away, and you can forget the massive bike event a short distance away.

Sturgis Bike Week

The Sturgis Bike Week, or to give it its correct title, the Black Hills Motor Classic, is probably the biggest bike event in American, after Daytona. It did not, however, grow out of racing, so it is probably the biggest bike gathering of its type anywhere. In fact, the town has the honorary title of Bike City USA – for the month of August at least!

It's certainly the best known the world over and is seen by many as a predominantly Harley gathering; but when you look back at its origins, Harley-Davidson was not part of the initial scene at all. A local Indian motorcycle dealer by the name of Clarence J.C. 'Pappy' Hoel started it in 1938, along with friends from his motorcycle club, the Jack Pine Gypsies.

Motorcycle stunts predominated in those early meetings, Hoel being something of a performer himself. These ran alongside AMA-sanctioned races, which were part of the reformed Gypsy Tour mentioned earlier. From this evolved several small events all around the locality, which all intertwined and became part of a ten-day happening that now sees 400,000-plus people arriving in the romantic Black Hills of South Dakota.

Although given the generic name of Sturgis, while the event was on, the actual town of Sturgis was the domain of the AMA breed of clean-cut racing types. Down at Deadwood, a small town 10 miles (16km) away, a different type of visitor was setting camp, and the town became the preserve of outlaw gangs: eventually, however, the two factions met and merged as they sought out things to do and places to visit. What with Harley-Davidson basing itself in Rapid City just down the road, the whole scene started to expand, with riders doing the rounds of

COLE'S
MUSEUM
The New Dodge
OFFICIAL VEHICLE
Sturgis '98
CENEX
STOP
NO U TURNS
5.99 EA OR 2

Smiley's Tap is another of the perennial favourites at Daytona.

The alarmingly named Road Kill Café at Sturgis, next door to the while-U-wait tattoo and body-piercing parlour.

towns in the area until the event became known as the Sturgis Rally.

In the beginning, however, it was very much a local event, quite a small biker meeting in the summer and almost an invitation-only gathering. It remained this way until the end of the 1970s, when a major article in the bike press gave it a glowing report: TV cottoned on, featured Sturgis and the rest is history with thousands flocking to be part of it, which in turn encouraged Harley-Davidson to establish an official presence.

South Dakota has some of the best biking roads in the US, so this all dovetailed nicely. Harley put its efforts into displays at Rapid City where there was a big civic centre at its disposal. This gave riders a 30-mile (48-km) blast along the Interstate 90 from Sturgis and they began to make the event their own by running events and promotions.

But like Daytona there was a period of big trouble, which took some of the shine off the event. Again, like Daytona, things are much calmer these days, but the legacy remains of a heavy law enforcement presence that is firm but fair. Now, the greatest problem is overcrowding, pure and simple. In 1947, 400 people took part in the Black Hills – now it is more like 400,000, and people often have to stay out of town once the local accommodation and campsites have been filled. Like Daytona there is Main Street where bikes are parked five-deep down its half-mile length. Parties take place at virtually every campsite and there are unofficial competitions to go with the official hill climb that is still run as a throwback to those early days. Drag racing, as one might imagine, is rife, both organized and unofficial!

For many, however, it is the strange mix of outlaw bikers and weekend rebels that attracts them to Sturgis. While the lid

OPPOSITE: Cruising down Main Street is one-way, and done very slowly.

ABOVE: People, bikes and sideshows sum up the Sturgis scene.

ABOVE RIGHT: Just married: this well dressed couple at Sturgis have just tied the knot – on a bike!

remains firmly on, some of the partying, which continues day and night, is bordering on the obscene. Nothing seems to be too crazy or out of the way and people do not seem to mind cameras or videos recording their antics!

One thing is for sure: there is plenty for everyone, with 124 events recently recorded over the ten days. These are watched by visitors from 50 US states and 17 foreign countries, and take place day and night, emphasizing the fact that there is no time whatsoever to sleep.

Whatever the locals may privately think, Sturgis is big business and brings in a level of currency that would be hard to replace by means of normal tourism, no matter how picturesque the area is.

Bike City USA is safe for the foreseeable future and will predominantly remain the province of the fans of Harley-Davidson.

Laconia

Once upon a time the name Laconia stood alongside Daytona and Sturgis. Now efforts are being made to return the New England town to popularity with the Harley fraternity with the heavy promotion of Laconia Bike Week. First started in 1923, the event traditionally takes place during the beginning of June. It was based on racing (like Daytona), which had been actually been taking place since 1917. These were primitive races and events and it was not until the 1930s that AMA-sanctioned races began.

In 1990 the New Hampshire International Speedway Stadium was built near the town of Loudon, which confuses some, as the Laconia event is also sometimes known as Loudon. Similarly, the hill climb run at Mount Belknap also comes under the generic Laconia title. In the heart of the town (pop. 16,000) is Wiers Beach, a dilapidated run-down amusement mile on the shore of the lake. Here during Bike Week it becomes home to the bikers and unfortunately has been the scene of some quite serious rioting over the years. This stems back to the time of Hollister when it was hushed up, but it has led to the cancellation of the event at least once, in the 1960s. However, after some serious aggravation took place during the 1980s, the police took their revenge at the start of the 1990s without actually stopping the event.

Heavily restrictive traffic orders were implemented and deliberately caused

As night falls, it's business as usual on Main Street. This is one balmy evening in 1998.

If it gets too crowded in town, Sturgis is surrounded by undulating roads filled with other bikes.

massive traffic jams around the area. It became an offence to switch off engines, punishable by on-the-spot fines, with the result that engines overheated and died. Having proved their point, the authorities concentrated their efforts into organizing events to keep visitors happy and deter them from contemplating making trouble.

It has worked and now 200,000-plus make the trip to watch concerts, take part in competitions, attend bike shows and customs and watch the races in a peaceful, convivial atmosphere. Maybe Laconia can still sit alongside Sturgis and Daytona as one of the big three events of the year.

Just like every other big event, there's a great deal of variety at Sturgis. This line-up of custom bikes is presented by Mike la Fores and friends, while others prefer to take it easy on standard bikes. However, it's best not to inquire too deeply into 'No Panty Wednesday'.

OPPOSITE: Sturgis nightlife.

GOLD PAN
PIZZA
GOLD PAN PIZZA
OASIS
COURTESY PARKING
Big BaD BeRT

CUSTOMIZING

ABOVE:
'Shovel Trouble' was built in the UK.

OPPOSITE: A high-tech chopper. On the face of it, this bike is simply repeating a 30-year-old style. But look more closely, and innumerable details mark it out as a modern custom.

Motorcycling is very much an individual affair and for some a solitary pastime. Whatever it is, most riders usually have their own strong opinions and are not afraid to voice them or stand out from the crowd. While for some, the standard model Harley in the same trim as it left the factory is sufficient to satisfy their needs, others regard this as a starting point. These are the customizers who just cannot resist altering their bike's appearance or performance.

What does 'customizing' actually mean? The precise definition is to personalize, which is achieved by changing various components to suit the individual's taste. This may simply be a case of a new paint scheme, but for others it may be more radical, involving totally altering components, reducing the bike's weight, or making it fit them personally in an effort to make it more comfortable to ride. And for some reason, Harleys seem to be the most popular bikes for this purpose.

There are, of course, very practical reasons why this is so. Harleys, unlike many modern machines, can be worked on with ease. They are not as mechanically complicated as others and require fewer special tools for the job. Also, because they are built and designed with no end life, every part is replaceable

So parts are easily obtainable, and the cost of making changes, often a huge factor, can be spread over a long period of time. In other words, one can do small amounts at a time without having to strip the whole bike from top to bottom. Lastly, of course, there is the major fact that the engine is a magnificent lump of metal and the sound of the V-twin motor makes the whole effect complete. With most custom bikes, the top speed is immaterial, unless it is designed as a race bike; but even then there are sufficient tuning parts available to make it a viable part of the rebuild as well.

Custom bikes, from the excesses of the 1970s to today's exotic and varied creations, have become a permanent feature of the motorcycle scene. But the modern custom movement didn't get going until after the Second World War. As already discussed, the end of the war saw many GIs returning to America with the desire to ride a bike. Ex-army bikes were plentiful and cheap, but they were heavy and cumbersome. The cheapest and easiest way to make them faster was to cut the

weight, so certain parts were removed and consigned to the trash can.

It was this practical weight-saving exercise that led to the verb 'to chop' entering the biker's vocabulary – hence the 'chopper', used to refer to bikes with long raked out front ends, that were built out of machines and parts that had been chopped. Some maintain the name comes from chopping the frame to alter the steering geometry, but the end result is the same.

But like most things, chopping isn't as simple as it seems, as removing one part can make a bike appear lopsided or unbalanced; so owners began to build bikes from scratch rather than modifying a piece

ABOVE:
'Has Advantages', another modern chopper, with belt primary drive and twin front disc brakes, though the low seat/high bars look is an enduring one.

OPPOSITE: An unusual customized Harley Trike. Picture Courtesy of Flickr/El Caganer.

here or there as the fancy took them. This means *carte blanche* with surfaces, and ever more elaborate paintwork, murals and chrome began to make an appearance. Neither was it just cosmetic, with raked forks and high bars getting ever more radical year on year. Peter Fonda's classic chopper in *Easy Rider* inspired a generation of riders to do the same to their bikes, of whatever make, but during the 1970s there was a suspicion that it had all gone too far. Specialized custom shows encouraged people to build bikes that were designed for showing only, rather than riding: some custom bikes were becoming less and less practical for the road – more modern art than motorcycle – and some even had no internal engine.

While these initial efforts were nothing more than attempts to outdo the next guy, shrewd machinists and sprayers realized that a business opportunity had presented itself to make parts for others less skilled than themselves. As a result, a multimillion-dollar industry has resulted, allowing people to alter parts here and there by replacing components using the existing mountings.

Bolt-on customizing had arrived, and it is still an important part of the movement today, and of which Harley-Davidson takes full advantage. Early on, Harley realized it was its own bikes that were receiving all the attention. So it too began marketing its own performance and tuning parts as well as the more normal accessories like bags and screens. Now, the Harley-Davidson range of accessories and Screamin' Eagle tuning parts is vast, to rival any of the after-market manufacturers.

However, for the committed customizer, merely buying the part and bolting it on is not an option. They have to make as much of the bike as possible, in order to end up with something that is unique.

Another alternative, albeit an expensive one, is offered by the firms that will build a custom bike from scratch, to the buyer's own specification, when he chooses from various components in a catalogue. Some of these firms work on the sound principle that people do not want to buy a new bike, then discard half of it and spend thousands on new parts.

What they like to do instead is build something different from the start. Companies such as Titan, Big Dog Motorcycles, American Eagle, and American Legend build up the bike from scratch. Most use the S&S motor and frames built by various manufacturers, but put their own names on the tank. These tend to be classed as clones, as a quick uneducated glance would mark all of them down as Harley-Davidsons. In reality, not a single part will bear the Harley stamp.

ABOVE: A Ron Simms paint job, with skull and flames a favourite theme.

OPPOSITE: The ever-innovative Arlen Ness (left) with 'Convertible', one of the new breed of 'bodywork bikes'.

ACCEL
ACCEL
ACCEL
NAVY
USA

Not the real things, but replicas of the bikes Dennis Hopper and Peter Fonda rode in *Easy Rider*.

'Jesse James' is the name of this bike, the flames and staggered pipes giving a vivid impression of movement even when it's standing still.

ABOVE LEFT: Indian Larry's Custom Shop.

ABOVE: Arlen Ness' 'Ferrari Bike' was inspired by the Italian sports car.

LEFT: If your budget won't run to a professionally-built bike, buy a sticker instead.

OPPOSITE: Arlen Ness' son Cory has followed him into the custom business. This is his 'Y2K'.

Arlen Ness is undoubtedly the doyen of commercial customizing, having been involved in painting, modifying and building Harleys since the late 1960s. He now owns a multimillion-dollar business by providing ready-made components, building bikes and advertising them, in the process becoming the acknowledged guru of the custom scene. Based in California, his influence has spread the world over, his catalogue of parts and accessories being almost a wish list for those intending to personalize a bike, and he has provided the inspiration for many who have gone on to create their own masterpieces of motorcycle art.

Custom Sub-Groups

If one thing has distinguished the custom movement in recent years, it's the sheer diversity of bikes now being built. Given that the idea is to build unique, personalized machines, you might think that all of them are different. But within this diversity, it is still possible to find general categories of custom bike, in which the builder may or may not have intended to place it. What follows is a brief guide to the major types.

Choppers - The classic custom bike, typified by long raked front ends as exemplified by Peter Fonda's Captain

Y2K

America. The headstock of the frame is angled at 45 degrees and it normally has long extended forks or, if these are short, high handlebars. Most of these are hardtail frames, without rear suspension. For the well-informed these can be broken down into many sub-categories of chopper, depending on the style.

Full Dressers - These are the large touring bikes that have had just about every part of their exterior covered in extra lights and chrome, not to mention a few extra novelties. The name stems from the 1930s, when motorcycle tourists chose to 'dress' their Harleys or Indians with a screen and panniers, or maybe spotlights. Where paintwork can still be seen, a mural is usually depicted, which will often be lit by multi-coloured lights in the dark. You'll often find a stereo and sometimes a CB radio on the fairing, and nowadays even a personal phone or a satellite navigation system is likely to be there. Their forte is to parade during the hours of darkness, when their presence can hardly be ignored!

ABOVE: A hardtail Evo chopper, with Weber carburettor and neat pipe arrangement.

RIGHT: You need to be dedicated to ride in the classic apehanger position.

OPPOSITE: Something different. A mildly customized dresser: but the closer you look, the more you see. There is the paint job, the aftermarket front discs, as well as billet handlebar grips.

DANGER

Rat Bikes - These are different from other custom bikes (which are usually kept painstakingly clean and spotless, as if they had never travelled the road) and are at the opposite extreme, dirty workhorses that make a point of being just that and sport patches, bodges (replaced parts from non-motorcycle sources), and more grease and grime than paint and chrome. The typical rat bike shows signs of having covered thousand of miles with a hard-riding owner, who paradoxically may love his bike as much as anyone else.

But like any other customizers, rat bikers can go to extremes. Some owners now set out to encourage the process of decay, deliberately causing the bikes to rust, creating oil leaks and manufacturing the bodges before leaving the bike out in extremes of weather to create the desired effect and in their eyes making them mobile works of art in their own right.

Add a few dead critters, along with some faded ex-military luggage, and hey presto! you have a bike capable of winning a class at a custom show.

OPPOSITE: A gloriously tatty rat bike: one wonders how it manages to move at all!

RIGHT: A superbly decayed rat bike in all its flea-bitten glory. Is that a trailer behind?

Custom Bikes - This covers just about anything, but there are various classes within the category.

The term Lowrider speaks for itself: a bike that is long and low. This can be further sub-divided into a Taildragger, where the rear fender covers the rear wheel down to ground level. The Streamliner is where most of the bike is enclosed, save maybe the engine. The Classic, is where it is based on classic styling themes from the past, such as the Indian Chief, for example, or the Harley Knucklehead. A Street Dragster, is a long and low machine with a performance engine.

The Café Racer or Warbird - This is a bodywork or styling kit which is fitted to give the impression of a fighter plane. It is made by seat makers Corbin.

Multi-Engine speaks for itself. At least two or more engines are slotted into an extended frame, usually inspired by multi-engine drag-racing bikes – not always useable on the road! The Art Class is where the bike is made to look like a well-known object or represent an advertising campaign. Trikes are again a self-explanatory term, together with the Sidecar classes.

ABOVE LEFT: On some rat bikes, the less you can see of the bike, the prouder the owner will be.

LEFT: A radical Knucklehead chop from Milwaukee Iron.

OPPOSITE
ABOVE BOTH: The Corbin Warbird dramatically transforms the standard Sportster. This one has the classic Harley racing colours of black and orange.

BELOW: The Sportster as modern Café Racer – just to prove it can be done.

The Rat's Hole Custom Chopper Show

Having built a custom bike, most serious builders want to show it off to an admiring audience and pit it against other builders in organized competitions. There are many shows worldwide, but the best known and most prestigious takes place at Daytona each year during Bike Week. For some, the Rat's Hole is the only one worthwhile.

Karl Smith, a local Daytona businessman, known as Big Daddy Rat, started it in 1972. Smith acquired the name Rat when playing football at university in Ohio and the Big comes from his size. In fact he used to sell T-shirts.

The inaugural meeting was held on the parking lot of Johns Family Market. There was so much interest, that it is repeated each year on the last Saturday of Bike Week, when a modest entrance fee is charged. Its popularity grew so quickly that it was moved to the boardwalk in front of the ocean and that year 15,000 people came to view the bikes, which were vying for victory in 44 classes. Nowadays, over 400 bikes come from all over the world to compete, Big Daddy Rat still being one of the main judges.

Since 1986 it has also run at Sturgis and has been linked to various other custom shows throughout the world. The free newspaper, *Ratty Rag*, always gives visitors plenty of information on what is happening in the area and the low-down on the various bikes that scoop the prizes.

Despite being copied the world over, the Rat's Hole Custom Chopper Show is still the number one and one of the few to have Big Daddy's own custom trike, the 'Cheese Hauler' on display.

Vintage Harley

It is all very well looking to the future, but many prefer to look backwards to Harley's glorious past. Restoring and running bikes that may otherwise have ended up on the scrapheap is a good way to proceed. Most of these people, some as old as the bikes they run, others who had yet to be born when their own bike rolled off the production line, turn up at events run by the Antique Motorcycle Club of America, which was founded in 1954. These events allow people to display immaculately restored machines in the *concours d'élégance*, or show original bikes that have the patina of age.

There is a chance to ride the old bikes, discuss them, and discover the parts needed to bring a moribund Harley back to life or restore it to total original form. These swapmeets and flea markets have parts and accessories from all periods of time, as well as useful tools and books to help complete the projects.

It is fair to say that one can find just about any part ever made, providing one has the time and patience to look for them. However, many will be worn and need work: one alternative is to use the worn-out original as a pattern, and get a machine shop to make up a new part from scratch. Allied to this are opportunities for those wishing to race the bikes in various classes designed to cater for the classics. While some may be too rare to actually race and are only displayed from time to time, a greater number are ridden hard in anger. They also get run rather faster than when they left the factory, thanks to modern tuning knowledge, coupled with the use of modern lightweight materials used in the rebuilding of the motors.

Then there are the trials and motocross events, plus Short Track and Long Track

OPPOSITE: A soft and smooth streamliner with all the appearance of a blob of whipped cream.

LEFT: Despite being recently built, this bike is based around a supercharged Shovelhead motor. Leading axle forks and extended rear mudguard add to the long low look.

BELOW: Two more traditional chops, both Knucklehead-based. They are the sort of thing outlaws would have ridden in the late 1950s and '60s.

races run specifically for period bikes. Here the bikes run as they did in their prime, giving the rider and the spectator the chance to sample some real nostalgia. Many of those who indulge this vintage or classic passion are not interested in modern Harleys, despite the fact that even the latest range owes much to the past. Instead, they see themselves as saviours of bikes that need to be preserved for the future as a reminder of Harley's greatness and also as symbols of a gentler age.

Art, Culture and Collectibles

With such a fanatical following it was inevitable that books should be written about the marque, magazines produced, pictures drawn, and lots of photographs taken. To list all the books and magazines that cater specifically for the brand would take a volume in itself, likewise for the artists who illustrate the bikes.

Some of course stand alone, like Scott Jacobs, whose photo-realist style has made him a licensed seller of Harley-authorized pictures. Dave Mann, the cartoonist from *Easyrider* magazine, and French artist François Bruère are just two other people who are renowned for their portrayal of Harley-Davidson machinery in drawings, the latter having built some superb custom bikes as well.

Drawings, of course, can cover the skin, and tattoos are popular with the cognoscenti. The Harley emblem is favourite, along with the 'Ride to Live' slogan to match it. Straightforward body painting, leathers adorned with painted devices, as well as airbrushed helmets to match the bikes, are all desirable accessories at bike shows and rallies.

Another form of art that is proliferating is sculpture. Life-sized bikes are appearing made of wood, bronze,

LEFT: Wherever they appear, custom bikes attract interest, causing people to stop in their tracks, whether it is to marvel, gawp or jeer. The point is, they never go unnoticed.

BELOW: The chequered speed theme is striking on this modern Shovelhead custom.

OPPOSITE: Every single component of this bike is made of wood; however, it won't run, even if it looks as though it might.

HARLEYWOOD

scrap parts, resin, in fact you name it, it has been used. The bikes have either been made as separate items, or as part of a thematic presentation, as with the work of Mark Patrick.

For the less ambitious, there are countless model kits of Harley-Davidsons, either ready-assembled or in kit form. Then there are belt buckles, watches, clocks, glasses, mugs, the list goes on, not to mention the vast selection of other goodies and clothing marketed directly by Harley itself, who only spotted the earning potential of its name in the early 1980s.

By then it had lost millions, as many others had already made money out of selling products bearing the Harley-Davidson name. However, after a great deal of work in the courts of the world, it managed to get a grip on the situation, although it will never be totally in control.

Nowadays, the ardent fan can be literally dressed from head to toe in Harley clothes. They can bathe in Harley oils, cover themselves in Harley talc and deodorant, not to mention the odd splash of aftershave or perfume.

No longer do you need Harley grease under your fingernails to show you ride one. Moreover, at the end of the day, you can end your ride with a Harley beer before snuggling under (you've guessed it) a Harley duvet! Harley-Davidson's list of products and collectibles is far too long to cover here; suffice to say that its marketing team is making up some of the losses from the time its goods were not licensed.

So who buys all this stuff? Who are the Harley people?

Harley People

The popular image of a Harley rider is well known, but in reality it will never be easy to recognize one across a crowded room because they exist in every spectrum of society. Quite apart from patch-club members, there are women riders, judges, politicians, police officers, actors, pop stars and of course the average man in the street. If you'll forgive some stereotyping, what follows is an attempt to identify some categories. In reality, the boundaries are blurred and many people have a bit of two or more groups in their makeup.

If you wish to get away from Sturgis, you can always tour the countryside. This group is checking out deserted mining towns in South Dakota. Second from the right is photographer Garry Stuart, who was responsible for most of the pictures in this book.

ROCHFORD
POP. 25
ELEV. 5305

The One Percenters
Perhaps the strongest identification of all, they are the people who love being different, still dress in the style of the 1960s rider and feel they should act in the same way as the people responsible for Hollister.

Having said that, most of them keep themselves to themselves and fraternize only with their own kind. They ride heavily modified bikes and do not conform to society in general.

Stereotypes
Closely allied to the one percenters is the stereotype biker or, as some would call them, the cult bikers. They dress all in black, wear heavy Indian jewellery, have tattoos and behave as though they were real outlaws. In truth, most are hardworking family members who like to pose and escape at weekends. They ride pretty much standard bikes.

The Normal Rider
These form the bulk of Harley riders. They are often family people who work in normal jobs and run a Harley as a means of escaping at weekends. Their bikes are not always new, but they don't care as long as it is a Harley, to which they are unswervingly loyal. Not as heavily into the tattoo/jewellery image as the 'stereotypes', they are just as committed to riding Harley-Davidsons.

RUBs or Rubbies
RUB is short for Rich Urban Biker. This sector comes from the high-earning faction, many of whom have come recently to motorcycling and see it as a leisure activity, like hang-gliding or mountain-biking. They require a bike to match the new car each year and are prepared to spend whatever it takes. Money is not a problem and they ride for relaxation rather than escapism. Because of their spending power, this group, far more than any other, is now Harley's target audience.

Although some profess disdain for the Rubbies, their practice of changing bikes almost every year ensures a constant supply of low-mileage secondhand Harleys onto the market. They keep the public profile of Harley-Davidson high, not to mention making a substantial contribution to company profits.

The Rich and Famous
Harley-Davidson has had a long relationship with celebrity riders who have been keen to associate themselves with the image. It is product placement of the highest order, and a marketing man's dream. In reality, celebrities only ride spasmodically, having to fit riding time into busy schedules. For Harley-Davidson though, they remain one of the best advertisements it can have.

So there you have it, a short resumé of the Harley lifestyle with all its associations and events. It will go some way to explaining what makes these people tick, and why they buy and ride Harley-Davidsons. At the end of the day, however, every rider has his or her own personal agenda, and what matters is that it works for them. Ride Free!

RIGHT: Ten per cent of Harley riders are female – you want to argue?

OPPOSITE: Taildraggers: modern customs with plenty of nods to the past.

L315
EHO

MILITARY HARLEY-DAVIDSON

It was in April 1917 that America actually entered the First World War, although long before this it had been supporting the Allies with supplies and vehicles. Like all the major US bike manufacturers, Harley-Davidson had been fulfilling military contracts that came its way since 1915, so the mechanism was already in place to supply in volume the distinctive drab olive green machines to the armed forces.

The company had also learned valuable lessons concerning military requirements when its bikes, in both solo and sidecar trim, had been part of a home- based military operation. This took place in 1916 after the Mexican government had allowed the US to enter its territory to deal with the problems that it was encountering with bandits, who had been crossing the border and attacking and sometimes killing Americans in adjacent townships.

The force of 20,000 soldiers, under the command of General John 'Black Jack' Pershing, failed to capture the freedom fighter Pancho Villa and his men, but they did gain valuable experience in the use of motorized units, known as 'Mounted Infantry', or 'Motor Mobile Infantry', and they were used where heavier vehicles were unable to go because of the mountainous terrain. Pancho Villa himself was a motorcyclist, owning an Indian, but despite the failure of Pershing's campaign, motorcycles had proved their worth on the battlefield. There was a roughly fifty-fifty split between Harleys and Indians in the unit, with just a few Excelsior machines thrown in for good measure.

On the strength of this, and the huge orders which were beginning to come through from Europe at the outset of the First World War, Indian nearly bankrupted itself and its dealers by turning over virtually all its production to the military, and starving civilian showrooms. It also sold the military machines at almost cost, which may have been laudable for the war effort, but didn't do its cash flow any good.

Harley was rather more cautious, and it was not until the middle of 1917, after America had entered the war, that it actually turned 50 per cent of its production over to military machines. Many of these bikes were shipped abroad, either to the front lines or to be used at bases as supply vehicles, or as part of intelligence-gathering operations.

ABOVE: This 1917 military motorcycle with box sidecar was exported to Russia.

OPPOSITE LEFT: *The Enthusiast*, Harley-Davidson's in-house magazine, was happy to emphasize its part in the war effort. This 1942 cover shows a US Army-specification WLA.

RIGHT: This rare machine, a flathead big twin, was supplied to the British military (hence its sidecar position). Most of the Harleys bought by Britain during WWII were 45-ci (750-cc) WLAs, brought in when its own motorcycle factories were destroyed by the Luftwaffe.

The Enthusiast
A MAGAZINE FOR MOTORCYCLISTS
DECEMBER
1942

ABOVE: Another view of the British outfit, with spare wheel, large solo saddle and hand gearchange.

ABOVE RIGHT: The WLA, with its rifle scabbard to the right of the front wheel, documents box on the left, and siren mounted on the handlbars. The 'L' plate is genuine, for many servicemen learned to drive during the war.

RIGHT: The speedometer may have promised a full 120-mph (193km/h), but a fully laden WLA could barely manage half of that. How would Steve McQueen have fared if he had used this in *The Great Escape?*

A fully-equipped WLA. The changes from the civilian bike were extensive, unlike earlier military machines.

One problem that was readily apparent was the lack of training, not only of the riders, but also of those expected to maintain the bikes and keep them running, often in difficult conditions and sometimes with incorrect equipment. Training schools were eventually established by Thomas Butler, a government representative appointed to solve this problem, who liased with the various manufacturers meeting military contracts and formulated a common strategy. These 'Training and Servicing Facilities', as they were known, were most successful, and paved the way for bike and vehicle manufacturers to found their own service schools after the war. The Harley-Davidson school was set up by Joseph Ryan, and eventually trained about 300 people during the war, going on to train dealers afterwards.

In fact, Harley-Davidson did well out of the First World War. Quite apart from making sure to keep the home market supplied, as well as the military contracts, it was able to increase exports while most of the competition was overseas and out of the picture due to war. It even managed to bag a few military contracts from foreign forces long before it really supplied the US military in a big way. One of these was the Dutch army and the other, the Russian, which took delivery of gun-carrying outfits along with ambulance units. The latter carried the stretcher at saddle height, the patient benefiting from the soft ride provided by the decent sidecar springs.

Apart from special outfits like these, there were few differences between the specifications of civilian and military machines, and in fact both benefited at the time from the general improvements in Harleys. The V-twin Model 8 had progressed in leaps and bounds with better handling as a result of the rider being repositioned near the centre of the bike. Rider comfort had also improved, due to better springing, a lower seat height, and a more acceptable riding position that made it less tiring to ride

The motors themselves had been re-engineered, which resulted in more power: the capacity was also upped to 61ci (1000cc) to bring it in line with the opposition. Then there was a hub clutch, step starter – the list went on. In 1915 the latest version of the Model 11J appeared with 98 improvements. Now with a three-speed gearbox and dry multiplate clutch that could easily be adjusted, it had a proper gate on the gear lever. There were numerous engine improvements and it now gave 15bhp: this, coupled with the improved gearbox, made it popular as a sidecar hauler. With this in mind, the frame was strengthened around the head-stock and it was generally beefed up to cope. Perhaps more importantly, it was the first bike to be offered with an electric headlight as a standard fitment.

In 1916 the pedals as a means of starting disappeared totally, replaced by the kick-starter, and new forks appeared on the Model 16J, which had various styling alterations to make it better-looking as well as functional. Bigger fillers on the petrol tanks went with petrol taps below, while filters were fitted inside to help stop blockages in the carburettor due to poor filtration. These benefits were further enhanced a year later on the 17J with the introduction of a generator and battery system. This was just in time to make it the model that would take Harley to war, albeit with a reduction in bright nickel plating.

Some Harleys had already seen action, having gone to war with the Canadians, who declared war shortly after Great Britain. The riders, who were often the owners as well, soon found the weight to be a problem, compared to the smaller British-built machines in service alongside them. This was particularly noticeable when bikes needed to be pushed through the inevitable mud of battlegrounds. As a result, most were retained and used for long runs on better-surfaced roads, where reliability and comfort were more important than off-road capability.

The army made certain demands when it placed those first contracts to ensure that bikes became as universal as possible. This was to make rider training, such as it was, easier, and to facilitate the stocking and supply of spare parts. All bikes had to have a right-hand twistgrip, the gearshift lever on the left of the fuel

OPPOSITE: Some people collect racers, others scooters, but many riders are keen on military machines. These three riders from Holland have a trio of similar WLAs, originally used by military police. Note the rudimentary screen on the bike on the right: protection obviously came before asthetics!

BELOW: The power unit for the WLA (though this is actually a WLC, the Canadian-specification machine) was a detuned version of the long-running 45-ci (750-cc) side-valve V-twin. Lower compression, aluminium cylinder heads and a bash plate were the main changes from the civilian specification. From its role as the workhorse of Main Street, whether powering a bike or Servicar trike, this motor found its moment of glory in wartime.

MP
MP
MP

tank, with the lowest gear all the way forward. The clutch was to be foot-operated on the left, with the rear brake on the right. All fittings were to be of a standard size, which assisted with the number of tools required, as were the chain sizes for driving the bikes. The tyres were to measure 28 x 3-in (711 x 76-mm). Happily, the Harley-Davidson Model J met all the demands from the word go.

The US government ordered about 70,000 bikes in 1917 and this contract was split between Harley, Indian and Cleveland. Research suggests that Harley-Davidson supplied 26,486 bikes, and while it failed to make a massive profit from them, it fared far better than Indian, which sold at a near loss. It would seem that 7,000 of these Harleys were shipped to England and France to see active service as convoy escorts, in dispatch roles, scouting and intelligence work.

The models the military was using were mostly V-twins: the 18J with generator and battery, the 18F with magneto, and the 18E model, which was directly geared to the rear wheel. Due to the contract requirements, there was a large amount of interchangeability between components on these bikes. All models benefited from the new lubrication system to the inlet-valve mechanism and a new bearing for the clutch.

Of course, this contribution to the war effort was of benefit to Harley-Davidson, but it was nothing compared to the effect of Corporal Roy Holtz. In November 1918, in Germany, an army rider was pictured in the US papers riding a Harley outfit past captured German soldiers. Pictured with the caption, 'The first Yank to enter Germany', Corporal Holtz achieved fame as a result of actually having been captured the day before. Lost behind enemy lines on the last day of the war, he was released immediately after the Armistice was signed. But he was pictured heading back to meet the advance, not as was claimed at the time, heading it!

When the American War Department finally closed all its contracts for the supply of bikes at the beginning of 1919, it had 14,600 Harley-Davidsons still on its books, 300 of these being sidecars outfits. So ended Harley's first involvement in military transport, but it was not its last!

The war had done Harley a few favours and now in peacetime it was clearly the largest motorcycle producer in the US. Some of the smaller firms had disappeared altogether, having failed to survive in any shape or form. Others were in difficulty, having had no military contracts to sustain them and therefore had been unable to purchase raw materials to build bikes for a dwindling market.

Harley's J-model was sufficient to carry it through the immediate post-war

ABOVE: The definitive 1942 WLA. Note the high-clearance front mudguard to prevent mud clogging the wheels when riding off-road. These bikes were expected to do everything and if the going got sticky, there was always the shovel.

OPPOSITE: Many collectors of militaria like to dress up in period costume. This man and his bike look ready for anything.

A pristine WLA: however, once in the field, it wouldn't stay like that for long.

years, being suitable for police use and also as a civilian mount sold through its dealerships. This was to give Harley the breathing space it needed to develop more up-to-date models. In 1918 it invested a further $3.5 million in the factory, re-equipping it with the latest machinery and adopting the newest mass-production techniques.

The new Sport Twin of 1919 was quite unlike any previous Harley and bore a strong resemblance to the flat-twin British Douglas machines that had served so well during the war. The US military bought three for testing, seeing its potential as a lighter alternative to the traditional V-twin, although there is no record of how they fared or whether there were any repeat orders.

One spin-off from the war was that many Europeans, who had never previously seen a Harley, were brought into contact with them. They were able to buy up surplus bikes that weren't shipped home, forming the basis for Harley's export market. The Model J actually continued to be built until 1929, so spares, such as were required, were always readily available from the factory. The military's own spares stock was sufficient to keep it going anyway, but in 1932 it ordered a batch of 100 R models to supplement the bikes still on the fleet.

Two years later it took delivery of some VDS models, which were based on the V and VL bikes intended purely for the civilian market. These were fitted with the LT sidecar, but a few sidevalve RL models

ABOVE LEFT: A WLA and sidecar in desert rig. It would have needed that extra-large air cleaner to cope with the sand, though this was part of the standard WLA specification.

ABOVE: The mirror is a later fitment to this WLA, to help cope with modern traffic.

OPPOSITE: Matching military bikes, with civilian cousins.

ABOVE: The Canadian specification WLC had left-hand clutch and brake lever. After the war (in fact for about another ten years) it was still possible to buy surplus WLAs that had never actually been used and spares weren't a problem either. The spotlight is an extra.

RIGHT: The rifle scabbard was standard, with the gun stock poking out, ready for use. In the event of suddenly coming under enemy fire, riders were instructed to leap straight off the bike and grab the gun: there was to be no nonsense about stopping first!

in solo trim also came with a repeat order in 1936 for a few more. A year on, the WL had its first mention in the order books, but for only 50 machines.

Although in military terms business was slow, with just a few overseas contracts keeping things going, Harley was doing quite well. What with police machines and sales in general it was certainly doing better than Indian, while Excelsior, the third member of the Big Three, had long since gone to the wall.

But the situation was soon to change: even before war was again declared in Europe, the US had been convinced that it would be involved. But as had happened 20 years before, Harley-Davidson was supplying military contracts well before war was declared by the White House. Five thousand solo machines were shipped to

England, for example, after the Luftwaffe destroyed the Triumph factory at Coventry. Not to be caught out this time, Harley re-opened its service school, training mechanics and at the same time submitting bikes for the army's extensive testing programme. Harley itself had two test tracks at Fort Knox and Louisiana, where bikes were subjected to merciless endurance and engine fatigue tests.

Finally, after a great deal of thought, Harley produced what it considered to be the best possible machine for military use, the WLA, the A standing for Army. This was a militarized version of the 45-ci (750-cc) side-valve twin that had been selling well in the late 1930s in the normal civilian market. Its compression had been lowered and it now sported aluminium cylinder heads to prevent overheating problems caused by escort work. Larger bearings, a better oil pump, and a stronger gearbox and clutch were also part of the improvements to deal with low-speed convoy work. It was tuned to give power at lower revs at the expense of top-speed performance, which was not seen as important. Fitted with a special large air filter for the Linkert carburettor, this would enable it to deal with hostile running conditions. The tubular forks were lengthened by 2.75in (7cm) to help increase ground clearance, and comfort for the rider was ensured by a large pan saddle supported on a sprung pillar. There were crash bars to help protect it from damage and a bash plate to look after the underside of the engine and frame. In addition, there were various attachment points for different weapons. The standard package also included a fork-mounted scabbard for a gun and a rack for carrying items at the rear. This was capable of carrying a 40-lb (18-kg) field radio and two saddlebags full of other items, without breaking down too often.

ABOVE: There is a good set of lights on this bike, but those stationed in Britain had to comply with blackout regulations. Headlights had to be heavily shielded to prevent being spotted by night bombers. This, and the lack of street lighting, led to a high accident rate during World War II.

OPPOSITE
TOP LEFT: A neat personalized plate for this WLA.

TOP CENTRE: The extra-large air cleaner was a standard fitment.

TOP RIGHT: No lack of storage space.

BELOW: One estimate numbers 88,000 WLAs like this rolling out of Milwaukee, though the true figure is a mystery. An order for 11,000 was cancelled in 1945, when victory seemed just over the horizon.

There was also a WLC model, the C indicating Canada. This model had the clutch lever and brake on the left with the gearchange on the right as specified by requisition orders from the Canadian military. It also had a box on the front fender to carry miscellaneous items. The WSR also appeared and was a sidecar version for the Russian forces, which other military outfits also bought, finding it suitable for their needs.

Unlike the previous conflict, civilian sales virtually ceased as the war progressed, though bikes were built for the home market. Home-based forces got grey-painted bikes, as opposed to olive green, with SP painted on them to denote their function as Shore Patrol. Apart from these general-purpose machines, records show several 74-ci (1200-cc) UA models along with some 61-ci (1000-cc) E and ELC OHV bikes bought for military staff use.

But the WLA was Harley-Davidson's standard as well as most often produced military machine. The US Amy ordered an initial 421 of them in 1940, followed by two further orders totalling 2,451 bikes. In 1942 a slightly modified version was produced, remaining the stock army bike until 1945. It has been suggested that over 88,000 bikes were produced for the forces at Milwaukee during the war, but this is always going to be difficult to confirm, especially with the various versions on offer. In 1945 Harley had an order for over 11,000 bikes cancelled when the government sensed victory, which confuses matters even more.

Certainly, when the hostilities ceased, there were several thousand brand-new bikes left with enough parts to build an estimated 30,000 more. This is confirmed by the fact that the bikes were still being sold into the mid-1950s at rock-bottom prices, complete with all military fittings! The price was set at $450, for which you

could have a ten-year-old bike that had never been used.

But while the WLA was well known as America's ubiquitous military machine, it was almost eclipsed by something very different – the XA. This was a horizontally-opposed twin, and an almost exact copy of the equivalent BMW. It was Harley's second departure from the V-twin and again came as a result of its military involvement: despite some reluctance from Harley-Davidson, the top brass was convinced that a shaft-drive flat-twin would make the perfect military motorcycle.

The company's answer, after purchasing a BMW for evaluation, was the XA, which produced 23bhp at 4,600 rpm and had a low 5.7:1 compression ratio to allow it to function in desert conditions. With its cylinders sticking out into the air for additional cooling, it was hoped that it would fare better than a V-twin or single in hot weather. It also featured a wet sump to tackle the heat problem and was estimated to run nearly 100 degrees cooler than a V-twin motor.

Other notable design features on this

The XA. Built at the US military's insistence, and heavily based on a pre-war BMW, the XA is the only Harley-Davidson (to date) with shaft drive. It was more advanced than the WLA, but more expensive as well. Harley-Davidson did produce a civilianized prototype in 1946, but nothing came of it.

ABOVE: The XA used the standard springer fork.

RIGHT: The power unit was a mildly tuned side-valve flat twin of 23bhp. It ran cooler than the WLA, due to those well-aired cylinders. One thousand XAs were produced, a tiny number by wartime mass-production standards.

model included a hand clutch, foot gearchange, with a down-for-down and up-for-up shift pattern on the lever. Some were fitted with sidecars and Harley experimented with a drive to the sidecar wheel like the BMWs, although this was never put into production.

During testing, a few problems emerged, but not enough to stall the project: at one time there was talk of an order for 25,000 XAs, and a reduction of WLA orders. In the end, the army took 1,000 in 1943 and then no more. None of them actually saw active service, remaining in the US at military bases. Despite all the development work that had been put into the XA and the lessons learned about such things as shaft drive,

LEFT and ABOVE: The Military Police-specification WLA, with shrouded light sitting alongside the main (legal) headlight, and the siren beneath it. It was enough to strike terror into the bravest AWOL recruit, unless he happened to be riding a Knucklehead.

OPPOSITE: This is not a weapon, contrary to rumour, but a standard-issue baseball bat for MPs! Look closely and you'll see the gloves, confirming that it is a recreational implement.

MP
Liberator
MP

Harley never used any of this knowledge or engineering in post-war bikes, almost to the extent of pretending the XA had never happened.

Other prototypes and experiments tried during the Second World War included a sidecar with a driven wheel like the BMW outfits. This was a totally separate project to the XA and was called the Model XAS. It was tried at the insistence of the US government in 1942, but the right-hand mounted sidecar machine failed to get further than the prototype. There was a separate design study for a machine using a 45ci (750cc) engine with special forks, frame and sidecar, which again came to nothing.

Slightly more suitable were the various trike projects (see A–Z chapter) one being a Servicar with a shaft-drive Knucklehead motor and which rolled along on solid disc wheels. Designated the TA, it suffered from handling problems, lots of vibration, and leaked oil in copious amounts. These problems were rectified, and despite being approved by the army in 1941, it went no further and certainly not into production. Another undeveloped prototype was a Canadian mini-tank, powered by two Knucklehead motors mated together.

But while experiments were in process, the solo motorcycle continued to have a role in the war effort. While radios were more reliable than ever before, messages could still be intercepted and there was still no substitute for hand delivery, for which a motorcycle was invaribly the fastest means. Dispatch riders were also the best people to keep intelligence up to date in the constantly changing scene of the battleground. However, the role was still basically one of support: leading convoys was probably the closest the rider got to the front line along with DR work. This non-

OPPOSITE: These are the items no WLA rider could afford to be without: gas mask, waterproof boots, radio, canvas bucket, tyre pump, full toolkit, spares, bed roll, mess kit and instruction book. The radio alone weighed 40lb (18kg).

ABOVE: Not a hint of chrome anywhere, the WLA was functional rather than pretty.

combat role did not prevent a fairly high casualty rate; but this was caused by accidents rather than enemy gunfire.

It should also be remembered that it was not only in Europe that Harley WLAs performed their duty. North Africa was another, and they appeared throughout the Commonwealth as countries began to use them as security or shore patrols like the US. In fact the US Navy also had a fair number of them travelling on ships to provide transport when docking at the various ports.

It should also be remembered that the Japanese were still producing Harley copies under the Rikuo, or 'King of the Road' brand name, having taken over total control of the Japanese factory from Harley in the late 1930s. These machines could also be found as part of the war machine in both solo and three-wheeler trim. There was also a much rarer Kuro Hagare, or 'Black Iron' machine in use, which was also Harley-based.

Meanwhile, Harley-Davidson was given the US Army/Navy 'E' award for excellence in 1943 and 1945, in recognition of its significant contribution to the war effort. This was all very well, but it seems that the government had failed to honour the total cost of spares produced by the company, and Harley is reported to have only made a 10 per cent profit on each bike it sold to the military. But as before, there were spin-offs: many overseas riders saw Harleys for the first time, and many GIs learned to ride at Uncle Sam's expense, with the result that when they returned after the war, they were keen to have their own bikes.

Throughout the 1950s, military production was almost non-existent. In 1963, Harley gained an order from the military police and shore patrols, totalling 418 bikes. These were XLAs, basically Sportsters fitted with fibreglass panniers and a windshield, and were more use on base rather than out in the operations field. In fact, the XLA was the last true Harley-Davidson supplied for military use. No new contracts came out of the Vietnam war, and in any case the army still had plenty of old WLAs in stock.

But Harley's military story doesn't end here, as the company did land another contract as late as the mid-1990s. Ten years earlier it had purchased the world manufacturing rights to the Armstrong MT500. Armstrong was a British company which was making a trail-style military bike powered by a 30-ci (500-cc) Austrian Rotax engine.

This successful machine had followed on from the Can Am Bombardier 15-ci (250-cc) single that had been acquired by many armed forces throughout the world, between 1977 and 1985.

It had revolutionized the military market with its lightness and responsiveness and was capable of withstanding both off-road as well as on-road situations without a weight penalty. Needless to say, it was popular with riders due to its all-round ability.

The Armstrong machine provided more power, being double the size, and had taken over quite nicely from the Can Am and was selling well. Harley was to re-badge these as its own, so when in 1995 the British Ministry of Defence awarded a new contract, it was to Harley-Davidson.

A total of 1,450 bikes were ordered and all were delivered on time; but three years later the contract was cancelled, possibly because the use of bikes for anything other than convoy work was diminishing.

FAR LEFT: So that there was no confusion, gas and oil tanks were clearly labelled

LEFT: The metal plate below the speedometer listed essential information like oil type, spark plug gap, and recommended cruising speed.

BELOW: Crash bars were useful if you needed to dive for cover and leave the bike to its own devices.

OPPOSITE: The WLC model intended for Canada.

On the road, the Japanese Honda Pan Europeans, military variants of police bikes, provided the escort bikes and other Japanese manufacturers now sell their trail bikes and Quads to military buyers who only need the odd off-road machine, rather than standardized military bikes as in the past. So it appears, as Harley moves towards its 100th birthday, that it will no longer be selling military bikes in any quantity: hopefully, there will be no need for it to do so again.

CAL .30M
AMMUNITION BOX

POLICE BIKES

Today, when Harley-Davidson is one of the foremost suppliers of motorcycles to the police, it's hard to believe that, not so long ago, it lost much lucrative business in that area. Everyone remembers *CHiPs*, the famous TV series based on the California Highway Patrol, that achieved a huge following the world over. But who remembers what they rode? They were not in fact Harleys, but Kawasaki KXp bikes. Actually, *CHiPs* was shown after Harley-Davidson had lost much of its traditional police business in the home market, as a result of poor quality control. It took several years of hard work, building up quality and holding down costs before the company was able to persuade US police forces to buy its motorcycles again.

But years before this, Harley had dominated the police market in America, from the early days right up until the late 1960s. Even when it was receiving stiff opposition from Indian and Excelsior-Henderson (with their super-smooth four-cylinder bikes that the patrolmen liked so much), Harley still had the lion's share of the market and always had since bikes were first used for the purpose.

Although for many years it was thought that the first bikes had been in service in Pittsburgh in 1909, it now emerges through detailed research, that Detroit had motorcycle cops in 1907. These were Harley-mounted in a town dominated by Ford and General Motors and demonstrates that they were far more use in congested streets than their four-wheeled counterparts. At the end of the century, some forces began to show renewed interest in two-wheelers, and for exactly the same reason!

In 1911, San Francisco and Los Angeles are recorded as having Harley motorcycle patrols and in 1913 Cincinnati followed suit with a bike squad. These early bikes are likely to have been the Model 5, a belt drive, pedal-assisted, single-speed machine of 30ci (500cc). Top speed was in the region of 45mph (72km/h), which was quite an achievement over basic rough roads more used to horse traffic. It is, of course, very easy to draw the analogy between the horse-mounted sheriff and the motorcycle-mounted officer who always worked alone. The difference was that the lawman was now more often chasing speeding cars than cattle rustlers. Tickets were handed out with a set lecture

ABOVE: Through the 1950s and '60s, Harley-Davidson was *the* US police motorcycle. The company was able to rely on police business, not only through patriotic loyalty, but also because no other bike of the time suited American conditions so well.

OPPOSITE: A magnificent 1938 UH side-valve sidecar outfit. Flashing lights and paint scheme are the only visible changes to the machine, though some details were different as well.

designed to make a person feel suitably chastened. The legend of the bike-mounted traffic policeman arose from the very outset of motorized transport, when he grew to be feared, even hated at times; but people also developed a sneaking regard for him as well, and of all the other police jobs, perhaps the motorcycle cop has retained an element of glamour to this day.

The work was also dangerous, and not only due to criminals: roads were poor and protective clothing minimal. But the cops were a highly visible deterrent to bad driving, and did a good job in reducing fatal accidents, a point Harley was quick to use as positive publicity. The publicity angle has been an important one ever since, and explains the cut-throat competition between American bike manufacturers to secure police business. Quite apart from the lucrative sales, bikes chosen by the police were seen as tough, dependable and speedy, and always on the side of the good guys.

However, it would be some years before Harley-Davidson set up a department dealing specifically with police contracts. It would be longer still before the bikes and the options available were any different from those on offer to the general public. However, sidecars were useful for police business, and after the First World War, the ex-military 61-ci (1000-cc) Model J (in both solo and sidecar trim) proved ideal for traffic enforcement and was sold on to the police to use up stocks no longer required by the armed forces. But in prohibition days, America was suffering a crime wave in which bootleggers on more powerful 74-ci (1200-cc) Harleys and Indians could outrun any lawman on a 61: the police had no option but to follow suit.

It was not until 1926 that Harley recognized the need to have a separate department to deal with police sales and

OPPOSITE: By the late 1950s, police bikes were more specialized. This bike started out as a touring specification Duo Glide with screen, spotlights and large rigid panniers. A police radio, fire extinguisher, flashing lights, siren and crash bars have been added. There were now painted wheel rims (not chrome), to save public money.

ABOVE: The standard Panhead V-twin, low-revving and reliable.

ABOVE RIGHT: A rider's eye view: the screen was essential for all-year-round use.

opened one up to liase with the 1,400 police departments which by then were using its bikes. It also launched an advertising campaign to capitalize on this huge success, 'Harley-Davidson will curb this traffic slaughter', referring to its huge take-up by police, compared to opposition like Indian. It worked and by the end of the 1920s 3,000 police departments were Harley-mounted!

Police riders themselves received no formal training, but were only required to prove that they could ride before joining up, and the'official' equipment was still almost non-existent. They were expected to work at all hours of the day and night and in all weathers. They had to be mechanics, paramedics and crime busters and be able to survive on their own in remote areas with little outside help. Radios had begun to appear in police cars in the late 1920s, but it took another ten years for some early unreliable units to appear on bikes. Until that time, the risks to the individual were high and the rider had to use public phone boxes to ring in every hour to give and receive relevant information concerning accidents and crimes.

Likewise, the relationship between police and dealers, who were expected to service these contracts, was not all that rosy. Although initial publicity was good, dealing thereafter with service contracts on tight budgets proved tough work, despite the bikes being almost identical to the standard civilian machines they were also selling. Most police bikes, however, had at least a tank-mounted speedometer driven off the rear wheel for enforcement use. But items like red lights, spotlights and klaxon horns were fitted by the riders themselves, often at their own expense. Even the bags containing report forms were often supplied by riders themselves and it was not until the R-model that the bikes came ready equipped for police.

Although the early 1930s was a generally bad time for Harley, police business continued to expand, while in 1932 the new Servicar provided an additional outlet. It quickly found favour with police departments, especially as a means of enforcing parking, thus helping sales figures which it would continue to do until the 1970s.

Harley realized its dependence on the increasing importance of police work at a time when civilian sales were slumping, and some specific pieces of police equipment were introduced: a police speedometer, a front-wheel siren, first aid kit and fire extinguisher. Then there was the Burgess muffler (which actually appeared in 1931) to direct exhaust gases downward, vital in the heavy traffic in which police riders often found themselves. Later safety guards and pursuit lamps were added as options.

By 1935 these options had been rationalized into two groups from which the police department could chose. The basic package included a safety guard, jiffy stand, 100-mile (161-km) maximum lighted speedometer and rear-wheel siren. The de luxe option added a steering damper, ride control, stop light, luggage carrier, first aid kit, fire extinguisher and pursuit lights.

There was also a 'Safety Silver' colour scheme to make the bike more visible at night, though it is fair to say that there was no standardization of police colours and markings at this time, each force doing its own thing as it does now; but the take-up on this option was not high. Uniforms varied from state to state, though most consisted of boots, breeches, jacket and caps.

What of the bikes themselves? Many departments had moved across to the JD-model for its extra speed and famous reliability, the California Highway Patrol being just one of note. It was, in fact, one of the first to switch to the twin headlight, side-valve Model D in 1930, and three years later designated the Harley as its official police bike throughout the state.

However, the later side-valve 74-ci (1200-cc) VL, with its early vibration problems, lost police sales.

In 1935, apart from new colour schemes, Harley was introducing a lighter feel to its bikes, achieved by relocating the tool box from the front fork to the main frame. Better brakes were also emphasized in advertising, achieved by fitting harder linings and using carburized brake drums.

Lights also received a makeover with a 'beehive'-style rear lamp, making it visible from either side as well as from the back. At the front was a pre-focused headlight, which concentrated the beam on the road ahead and allowed higher speeds at night, helping to cut accidents.

The factory was also offering a service that allowed older bikes to be returned and updated, effectively giving the forces a later model which was a cheaper option than buying new. (In the last few years, restorers have seen this as something of a problem when they realize they have incorrect or later parts on bikes they have purchased from a particular year.)

OPPOSITE: Caught in the act: 'May I see your driver's licence, sir?'

RIGHT: Police departments were long-time users of the Servicar, finding it useful for parking enforcement. One method was for an officer to ride past a line of parked cars with a piece of chalk on a long stick, when he would mark the cars illegally parked, leaving it to the meter maids to slap on the tickets. A Servicar could also carry a good amount of equipment in its box (or two or three hefty officers) but was still smaller and handier than a full-size squad car. It was not a pursuit vehicle, however.

In Milwaukee, the police department has been testing police-specification Buells, called Defenders. Based on the civilian Thunderbolt sports tourer, these have police equipment built into the panniers.

Now there *is* a pursuit vehicle. In town, the Buell Cyclone is fast and nimble enough to act as chaser and get to situations quickly. Note the blue lights peeking out from the headlamp fairing.

ETOWAH
FOR SALE

OPPOSITE: Like most police vehicles, the Servicars were sold off once past their prime. However, this has been given a new lease of life, having been converted into an ice cream van.

BELOW and RIGHT: The police were responsible for keeping the Servicar in production long after civilian orders dried up, and it continued to do so up to 1973. There were various updates over the years, such as three-wheel brakes with electric start later on. The Servicar was also fitted with a radio before the bikes, because it had available space (you can see the handset on this late model). It still sported a hand gearchange, but when you had three wheels to keep you upright, that hardly mattered.

The Knucklehead of 1936, though initially seen as a sports bike, soon found favour with the police, who were having to catch criminals using ever faster automobiles. That was also the first year the factory offered radios as a police option. The radio receiver cost $75 and the purchaser could state which radio frequency or wavelength they required, the whole unit being mounted on the luggage rack. Two years later, the price had risen to $135, and it was now manufactured by RCA, rather than Harley-Davidson itself.

The Servicar was still selling well, helped by updates such as a constant mesh gearbox, a strengthened rear axle, and brakes on all three wheels. It was also restyled, and by 1939 the box at the rear came in a variety of sizes. But despite all this, sales had been dwindling as some police departments sold off bikes in favour of cars. The reasons were many, but some felt that the car carried more equipment and was not that much slower; neither was there the massive price differential there once had been. Nevertheless, during the early 1940s, 3,500 police departments were using Harleys, with Indian on the slippery slope. It looked as though Harley-Davidson would have what police business was left all to itself.

During the war, Harley-Davidson cut

back its production for the civilian market to concentrate on police and military needs. So although the civilian market was virtually moribund, police sales and service were the means of Harley dealers staying in business. In 1942 and '43, production peaked at 29,000 bikes, the vast majority going to police and military contracts. Meanwhile, police work had undergone a change during wartime, with many forces carrying out duties that can best be described as paramilitary. Patrols of locations deemed to be sensitive were carried out by police officers, sometimes with Thompson sub-machine guns as part of their equipment, though never listed as a factory option, it has to be said.

Some unusual problems did surface. In 1942 all the American speed limits were reduced to conserve fuel, but bikes were wearing out cylinders in 6 –800 miles, despite the lower speeds. It turned out that the recently introduced radios required the engine to be kept almost constantly running, to maintain the battery charge and power it. Also, the nickel-alloy Harley-Davidson had used for plating cylinders, was now in short supply, and it had adopted a silicon substitute that was not up to the task, hence the high rate of wear. It tried chrome-plating the bores, which did the trick, and cylinder life increased to a rather more acceptable 20,000 miles or so.

This was fortunate, as the workload had increased, with many additional officers being employed for all the extra duty and patrol work. The bikes remained largely unchanged throughout this period, and Harley reduced the amount of chrome and brightwork by painting over these areas instead.

Around 1947, Harley resumed full-scale production of its bikes and began to enjoy something of a post-war boom. Thousands of returning servicemen wanted to buy bikes, having ridden them during their military service. They had enjoyed the experience, and wished to continue riding for pleasure. But the technical advances which had been achieved now benefited police riders as much as the emergent civilians; the Hydra-Glide fork, rubber-mounted handlebars, sprung seat post and 5 x 16-in (127 x 406-mm) tyres, all gave a more comfortable ride to officers facing a long hard day in the saddle.

Also welcome was the first of the two-way radios developed by RCA. They were first used on the Servicar, which had all the necessary space to house the hardware required. By 1948 the units had been trimmed to a compact 27lb (12kg) and were offered for solo use. They were fitted to the carrier at the back of the bike, taking up space once occupied by saddlebags. On the

ABOVE: On the look-out at Daytona. The FXRP was Harley's police bike for the 1990s, faster and more nimble than the traditional Electra Glide-based machines.

OPPOSITE: On duty at Sturgis. The police contingent at premier bike events, such as Sturgis and Daytona, are well aware that Harley-mounted officers will receive more co-operation and respect from the public.

POLICE
POLICE
STURGIS
POLICE
STURGIS
POLICE
61
60
Harley-Davidson
Harley-Davidson

SHERIFF
SHERIFF
SHERIFF
223
SHERIFF

OPPOSITE: Deterrents don't get more visible than this. This sheriff is based in Daytona.

ABOVE: Something to show the guys back home – speeding tickets, courtesy of the Daytona police.

RIGHT: A late-model Electra Glide: some forces still opted for the traditional police mount.

left was the receiver and on the right the transmitter. An aerial completed the rear equipment, but at the front was a speaker box mounted on the handlebars. Adjustable for volume and squelch, it sat right in front of the rider. The microphone came from the rear units and clipped onto the back end of the gas tank, so was easily accessible while on the move.

The problem was that while the officer now had two-way communication, the radios took all the available electrical power as well as storage space. The bike still needed to be kept at high revs to power it, which of course meant that it was virtually inaudible thanks to the engine noise drowning out the primitive speaker. It would not be until 12-volt systems came into being that the problem would really be solved, but at least the engines could now take the abuse, Harley having reverted to decent peacetime materials. Other police radios, including one from Motorola, soon followed, though the additional expense meant that many forces preferred their riders to carry on calling in from public call boxes. It wasn't until the 1950s that the introduction of transistors made radios lighter, more reliable, cheaper and less power-hungry.

By the 1950s Harley was enjoying something of a post-war boom and they were to be good years for the Milwaukee concern. In 1953 Harley lost its major competitor in the police market when Indian finally went to the wall. This left Harley as the only American motorcycle manufacturer, and certainly the only police supplier. A few forces did manage to keep their Indians going for another ten years before finally switching to Harleys, but most adopted them straight away as they required the back-up service.

But while there was now no home-based competition, smaller, lighter and cheaper Japanese bikes had begun to make an appearance on the general market and were of course later to have some bearing and effect on supply to the police.

Meanwhile, Harley-Davidson at long last offered a foot gearchange option, though oddly enough many police riders still specified the hand change, still available until the mid-1970s, which left hands free for traffic regulation and signalling. There were also new options on offer for the 74-ci (1200-cc) Panhead and these included high- and low-compression motors to cope with town use or the open road. The latter came with a lower ratio gearbox. Other options available were chromed safety bars, wheel rims and muffler, which all cost extra, as did the screen, which came in three colour options. Sirens were still driven off the rear wheel by a friction device, which occasionally failed to deal with new tyres that expanded as a result of heat. This caused a few of the units to self-destruct and many riders found it safer and quieter to have them behind rather than in front of them.

Surprisingly, the radio also remained on the options lists and was not standard, so in some US states tooled pannier bags were still seen attached to the rear carrier. Motorola had in fact introduced a smaller more compact unit specifically designed for a bike in 1957, instead of one adapted from a car. This was called the Dispatcher Radiophone and took very little power to keep going with a clarity not previously experienced. Designed to mount on the handlebars, it moved rearward in 1958 when the Duo Glide was introduced as a police motorcycle. This was of particular interest to police as it offered comfort levels as yet unexperienced, as well as increased safety features like hydraulic rear brakes. In reality, it was the venerable FLH-model, fitted with a rear suspension system. This was in response to claims of the rear carriers splitting due to vibration,

ABOVE: Close to home: This FXRP was spotted outside the Juneau Avenue works in Milwaukee.

OPPOSITE: Electra Glides of the Milwaukee Police Department – hardly a force to buy Hondas!

NO PARKING
POLIC

Harley's 95th-anniversary celebrations in Milwaukee saw a parade of police from all over the USA.

ABOVE: On patrol in Sturgis.

ABOVE RIGHT: No force is better known than CHiP, the California Highway Patrol, with its strong association with Harley-Davidson. It has been a loyal customer since 1933, so it was a big shock when it decided to switch to Kawasakis in 1975, a move which influenced many other forces to consider doing the same. Oddly enough, *CHiPs*, the hit TV series, featured patrolmen riding Kawasakis, though the public perception has long been that Californian patrolmen ride Harleys. This has turned out to be true since Harley-Davidson won back this prestigious contract.

especially when the weight of a radio was added. The lack of springing also caused some of these to fail on a regular basis, despite the new radio technology. There was also the fact that the FLH had a stronger transmission and clutch, along with an optional larger oil tank, all helping to keep it on the road longer and reduce mechanical breakdown. The weight had risen by 50lb (23kg) as result, but it was a small price to pay compared to all the other advantages.

Along with increased bike safety features, rider safety was now regarded as paramount: riders now got fibre helmets to replace the garrison-style caps that they had used before, and waterproofs were also beginning to appear in a uniform style. Rider training had also been taken on board with many forces organizing proper programmes to equip their riders to deal with all the likely hazards they would encounter on the road, like railroad tracks, cobblestones and errant drivers. Instruction was given on both road and dirt-riding techniques, especially in country areas. Low-speed as well as high-speed riding was also part of the equation, riders needing to be able to control their bikes in heavy traffic. In addition, of course, riders were required to complete daily safety checks of their bikes and maintain them in a clean condition.

Out of this grew many display teams, which allowed riders to demonstrate their prowess in front of large audiences. While they were representing their various forces in the often very impressive displays, they were at the same time providing much valuable publicity for Harley-Davidson.

But the 1960s were to have as much impact on the police motorcycle and ordinary markets as they did on society in general. 'Times they are a-changing', sang Bob Dylan and so it was with Harley, especially in 1969 when it was taken over by American Machine and Foundry (AMF). By then, police Harleys, like their civilian counterparts, had already gained 12-volt electrics and electric starting, both big leaps forward for long-suffering

SPD
STURGIS
THE
KEY CITY OF
THE BLACK HILLS
SD

OPPOSITE: Relaxed and in control: an officer of the Sturgis Police Department.

ABOVE: It's not only city police who ride Harleys: this FXRP is run by the San Francisco Airport Police.

motorcycle cops, who no longer had to jump on a kickstarter several times a day, or worry about a marginal electrical system keeping up with the demands of the radio.

The electric start had first appeared a year earlier on the Servicar, which was still going strong, in spite of the police being the only customers for the little workhorse. The change meant that a new primary casing had to be designed to deal with the starter motor drive, which in turn required the oil tank to move to the other side of the solo machine, which required a new frame in which to house it all. As a result, the weight of the bike rose yet again and new uprated brakes were therefore needed. Changes cost money and the price of the bike rose to the point that it was close to being double the cost of a car, which did not endear it to the purchasing departments.

Despite a power increase for the Shovelhead engine, police Harleys were finding themselves increasingly outclassed on the road by faster cars and bikes. In fact, the classic chase scenario grew less common as there was increased reliance on radio and radar to catch miscreants and discourage speeders. Harley-Davidson no longer fitted radios at the factory, leaving the market open for Motorola and others to meet the demand.

The Servicar received the12-volt alternator system as well, though there were still many early bikes in service. As before, a conversion kit was offered so that departments could update their Harley-Davidsons rather than buy new. The Servicar got its final update in 1967, when the box was made of fibreglass by the Tomahawk Company and incorporated the fenders into the moulding. In this form, the machine remained in production until 1973 when virtually all the orders for it dried up and it was no longer viable to make the hybrid solely for the police.

Meanwhile, as well as proper training and improved equipment, riders were now given extra pay in the form of a hazard allowance. Some claimed this was more for fixing the bikes, as the standard of quality control had declined and the machines were being stretched beyond their capabilities in the modern traffic conditions.

Certainly, Harley seemed to be increasingly indifferent to the needs of its police customer base, which had kept it going through some lean times in the past. This was a serious mistake and would take it a long time to repair the damage caused by lost sales. The Japanese were moving in on Harley's traditional market, building bigger bikes, while other manufacturers, who had a share of their own home police markets, were looking to expand there as well.

By this time, the California Highway Patrol had established itself as the most thorough tester and evaluator of police vehicles, so many other forces and agencies took a lead from them. If CHiP thought it was OK, it was good enough for them and it saved costs on doing their own trials.

So when they began looking at the Moto Guzzi 850 and Honda 750 police bikes in 1972, other departments sat up and took notice. But it was Kawasaki that secured the CHiP contract in 1975, when the department ordered 130 of its 55-ci (900-cc) police specials. The reason was very simple: Kawasaki had asked the officers who rode the bikes exactly what they expected of an official bike, as opposed to one for pleasure, and noted the replies; it then built a bike to correspond with those needs. It was faster, smaller and lighter than the equivalent Harley, easier to ride and faster to accelerate. The four-cylinder engine was smooth, and was mated to a five-speed gearbox. It ran longer between maintenance checks, broke down less often and spent less time off the road waiting for parts. So effective was the Kawasaki that it stole what few US police sales the Moto Guzzi had achieved. To show how serious it was about the police market, Kawasaki even set up a factory in Lincoln, Nebraska to manufacture the KZp 1000, as it later became. Harley-Davidson couldn't even accuse the Kawasaki-mounted officers of being unpatriotic!

It should also be borne in mind that, since 1933, CHiP had continued to have Harley as its official police motorcycle and its defection served to open the floodgates to other manufacturers. The official tenders now specified items that at that time deliberately excluded Harley bikes, so disenchanted were some authorities. Towards the end of the 1970s, Harley didn't even bother to offer bikes for the annual police evaluation tests, as some forces had stopped even looking at them as early as 1975.

But it would be wrong to infer that Harley did absolutely nothing for the police market during this period. In 1971 it had introduced the FLH Electra Glide with a 74-ci (1200-cc) high-compression engine and 10-inch (25.4-cm) front disc brakes, although the latter were found to be seriously lacking when under pressure. Electric sirens began to replace the wheel-driven mechanical items, and in 1978 an 80-ci (1340-cc) V-twin replaced the existing engine. This, however, still had an unacceptable level of vibration and the weight had increased even more. The suspension was not good and riders found the ergonomics totally unacceptable. Rider fatigue, poor handlebars, switches, operating of the clutch and gears were items adversely mentioned by the test riders. In the stability category they criticized it because it grounded out too easily and the suspension didn't cope too well. The engine was quoted as lacking in speed and power, to mention just a few of the complaints.

Not only that, but some forces were again having second thoughts about using motorcycles of any description: employment legislation was making them twitchy concerning the risks to which riders were being exposed. The effectiveness of the bike always relied on the skill of the rider, and not everyone had these skills. And patrol cars were cheaper than ever, forcing motorcycles into an ever more specialized role. Something had to change.

Change it did, for in the early 1980s, a management buyout caused a new attitude to emerge from Milwaukee, and one of the things that changed was the attitude to police business. Once again, the company made a real effort to obtain police sales, but it promised to be an uphill struggle. Whenever a Harley was tested against the big Kawasaki, the 1340-cc Harley came off worse in almost every category. Harley's police sales continued to decline.

It was probably the introduction of the Evolution engine in 1984 that was really the turnaround for Harley. After seven years at the drawing board, the rubber-mounted, belt drive, five-speed powerplant was radical indeed by Harley standards, but it was just what was needed to win back police business. A year later, Harley launched the FXRP and began to actively court police buyers again, which was to be a hard struggle with many having abandoned bikes altogether. There was also an east/west divide, with the Western police forces favouring Kawasakis; but Harley was at least making some progress in recapturing a little of its market share.

This was helped somewhat by the increasing amount of congestion on the roads, especially in towns and cities, which hampered response times to calls at a time when accountability was becoming a real issue. The only real solution was to revert to motorcycles.

Harley was able to climb back onto the evaluation lists, although it still came in for some serious criticism. But this at least pressurized it into dealing with complaints. Gradually, forces recapitulated and the familar sight of a traffic cop on a Harley reappeared on the scene. To achieve this, the company had had to offer deals that no doubt lost money. But it was necessary to regain a toehold in the market, and convince police buyers that the bad old days were finally over.

In 1991 the FXRP was substantially updated with new frame, forks, engine-mounting system, instruments, carburettor, seat, and starter: in fact the list ran to over 300 changes. Sitting behind the familiar fibreglass fairing was yet another level of sophistication with regard to law enforcement items. Now, a radar gun was part of the kit, as well as a PA system and improved emergency lights.

Three years later the bike was discontinued and the FLHTP appeared, with adjustable air suspension at both ends, complete with anti-dive on the front forks. Rider comfort was paramount and Harley was aware of this, realizing that if it wanted to continue winning back contracts it would have to meet this need and get the riders on its side. That is why items like the footboards became adjustable, the control layout became easier to read and everything was more accessible. The range of the bikes was improved with bigger gas tanks and the electrics were uprated. In short, Harley had listened to the needs of the professional rider, just as Kawasaki had nearly 20 years earlier. It had been a hard lesson, but Milwaukee had profited from its mistakes.

As the century drew to a close, Harley-Davidson introduced its first new engine for 15 years, the Twin Cam 88. Now increased in size to 88ci (1450cc), but still the distinctive 45-degree-twin design, it offered more power, was smoother-running and met all the relevant and increasingly stringent noise and pollution regulations. It was to revolutionize general market sales, and was attractive to the police market that was gradually improving. The Road King FLHPI and the Electra Glide FLHTPI were offered in fuel injection form.

In 1999, recognizing the almost cult status of police bikes among some civilian riders, Harley-Davidson offered a civilian 'Peace Officer Special' to enable officers to ride off-duty what they rode at work. In fact, it wasn't available to just anyone. It was a clever marketing strategy, but the FLHTCUI, in two-tone blue, was available only to those producing proof of employment with law enforcement agencies, to make sure it remained something special. As an end-of-century statement, it underlined the status of the police-specification Harley to the public at large. Although Harley-Davidson would never again have a monopoly on police business (BMW, for one, made inroads in the 1990s), its firm place in the market place seems at last assured.

OPPOSITE: Cruising by the Boothill Saloon in perfect formation, just to keep an eye on things.

RIGHT: Can you honestly imagine this officer happy on a Honda scooter?

THE BUELL STORY

On 19 February 1998, Erik Buell announced to the world that he had sold almost all his remaining interest in the Buell Motorcycle Company to his business partner, Harley-Davidson Inc, the latter having held a 49 per cent stake in the company since 1993. Buell was named chairman and chief technical officer, but his boss was Harley-Davidson, and he was effectively back to how he started out.

In one way or another, Harley-Davidson had long been associated with Buell the company: it had supplied engines from 1987, and in 1993 suggested to Erik that it might take a stake in his company. By the early 1990s, Harley had become a successful producer of cruisers, only one sector – albeit a large one – of the motorcycle market, and Buell was involved in something quite different: muscle bikes. From Erik's point of view, it was a chance for Buell to benefit from Harley-Davidson's huge resources and secure its future at a time that there was difficulty paying the monthly wage bill. So in early 1994 the new Buell Motorcycle Company was born, with 49 per cent of it owned by Harley-Davidson. For Buell, the association has worked well, with production rocketing from a few hundred a year to over 6,000.

ABOVE: Erik Buell, the man behind the bikes.

OPPOSITE: The Buell RR1000 Battletwin.

The Early Days

But none of this would have happened without Erik Buell himself, his hard work, and his dream. He had ridden his first motorcycle at the age of 12, and thereafter had been well and truly hooked. To pay for night school, he decided to take a job at a bike dealer's and worked his way up from trainee mechanic to service manager in a short space of time. At the same time, he began road racing at amateur level, and his dedication and talent soon established him in the realms of the professional. Racing a TZ750 in the Formula One and a Ducati 900SS in Superbike events he was soon up with the front runners and in 1978 recorded the fastest qualifying time for a newcomer in the Daytona 200. This will give some idea of his talent.

Despite his racing exploits, however, Erik still found time to study his textbooks and win a degree in engineering in 1979. He had worked for this with one intention in mind: to stay in America and to continue to work with motorcycles. This was at a time that there was only one major manufacturer: he went to work for Harley-Davidson as a chassis engineer.

It looked like the start of a successful career, and the still-young Buell managed to achieve several motorcycle-related component patents in his few years working at Milwaukee. But the dream of building his own race bike was still there, and in 1982 he left to do just that, forming the Buell Motorcycle Company in the process.

The first motorcycle that he designed and built was the RW750 racer in 1983. This was a 45-ci (750-cc) two-stroke, 'square-four', rotary-valve machine, designed to compete in the AMA Formula One road-racing class. His inspiration actually came from within Harley-Davidson, as the company had already produced drawings for a two-stroke four-cylinder engine of its own as a feasibility study, but had quickly abandoned the idea. It did, however, plant the seed of an idea in Erik Buell's mind.

The RW's engine actually came from England, being the same Barton-built motor that had powered David Essex's Silver Dream racer in the film of the same name. Buell also bought a frame from Barton, but was not satisfied with it and built his own chassis to house the motor. This was the first bike he built with his own name on the tank.

The prototype first hit the track at Pocono Speedway in the summer of 1983

Buell
Buell
AMERICAN MOTORCYCLES
RR 1000
BATTLETWIN
POWERED BY/HARLEY-DAVIDSON

ABOVE: An RR1200 of 1990 used the 74-ci (1203-cc) Sportster motor, suitably tuned.

RIGHT: Buells rapidly took to the tracks, which was fitting, given that their roots lay in racing.

in an AMA national race. Gradually, its builder (and rider) resolved what problems there were and by the the autumn the bike clocked 178mph (286.4km/h) on the Talladega track in Alabama. At this time, the Buell racing business was being funded by distributing parts brought in from England: exhaust pipes, Dymag wheels, Interstate leathers, to name but a few. He also sold items made by himself, with a view to keeping the cash coming in to allow development of the race bike to progress.

Buell carried on developing the RW750 throughout 1984 and a production version was released in the autumn of that year. He managed to sell just one RW750 to the American Machinists Racing Team, before the AMA announced that the following year

LEFT: But of course, Harley-Davidson had its own long racing history before Buell came on the scene. These are side-valve WR racers.

would be the last season of Formula One racing. Despite having the parts to build more RW750s, the project folded, so did the company and Buell lost everything, including his home.

At a stroke, the governing body of racing had effectively eliminated any future market for the RW750 by giving the premier slot to Superbike. But Buell viewed the rule change philosophically, and although tempted to go back and work for Harley, he changed direction and focused his attentions on building street bikes – machines that would be capable of making the transition to the track if the need arose.

The First Buells

Buell went back to work, this time with the goal of building the first world-class sports bike for the road, designed and built in the USA. Using his knowledge of what worked on the race track and his experience as a Harley-Davidson engineer, Buell went on to design his first candidate for the sports bike market: the RR1000.

This bike was remarkable in many ways, not the least of which was the fact that the project was started in September and the bike was up and running three months later. This was something of a record and one that Buell has yet to equal again. The RR used a Harley-Davidson XR1000 engine, which was mounted in a stiff, lightweight chassis. The frame was designed to allow rubber mounting of the powerplant, a system that has become a patented engineering trademark of Buell sports bikes. In addition, the design also utilized the engine as a planar stressed member of the frame.

Another Buell RR1000 first was the use of a rear suspension unit, working in tension, and mounted horizontally beneath the motor to allow a short wheelbase. This is another feature that can still be seen today on all the latest Buells. A total of 50 RR1000s were produced during the 1987–88 season, but Buell had already seen the new 74-ci (1203-cc) Evolution engine and saw it as a further step forward and another way of improving his bikes. Being mass-produced, it would be cheaper than the XR1000 V-twin, at the same time giving away little in power. He redesigned the chassis to accommodate the new engine and the RR1200 was introduced in 1988, with 65 produced for sale in 1989.

The same year, Buell also introduced the RS1200, a two-seat version of the RR model for riders who demanded both performance and two-up comfort. Just over 100 of these were produced through to 1990. The bikes also featured all-enveloping, streamlined bodywork, a fact which caused something of a stir at the time: never before had a bike powered by Harley-Davidson looked so modern or sporting.

Five-speed transmission was a new feature of the 1203-cc engine in 1991. Buell responded to revised engine-mounting points with further design improvements to the RS chassis. These bikes were the first production motorcycles

Buell

OPPOSITE: The limited-edition Buell XI, with 'upside-down' forks and that characteristic headlamp fairing.

BELOW: Buells have always been all-engine, bare-boned, minimalist machines.

RIGHT: RSS1200, a single-seat version of the RS1200.

BELOW RIGHT: Buells were the first Harley-powered bikes on which one could wear full leathers and a full-face helmet and not feel overdressed.

Buell

Buell
N537
CVV

Buell
AMERICAN MOTORCYCLES

OPPOSITE
ABOVE LEFT: The frame wraps around the engine: note the steering damper.

ABOVE RIGHT: Only a brave passenger would use that tiny pillion seat.

BELOW LEFT: The Lightning at rest.

BELOW RIGHT: There is a cutaway fuel tank on this show Buell to demonstrate the induction system.

ABOVE: The Uniplanar rubber mounting system quells vibration.

ABOVE RIGHT: A good view of the horizontally mounted shock absorber.

to use inverted front forks, stainless steel braided brake lines and a six-piston front-brake caliper. Manufacturing capability and quality control were also improved by the installation of a new composite and paint shop within the factory during this year. This allowed greater design flexibility and the capability to make changes quickly as well as aiding design work. A single-seat version of the RS1200 was then introduced for the 1992 model year. Dubbed the RSS1200, it won the enthusiastic approval of the motorcycle press on account of its lean, clean lines. Combined production of RSS and RS models through 1993 totalled 325.

But the company was still very small: Buell was employing just seven people at the time and needed a fresh injection of cash if it was to survive and develop. Hence the involvement of Harley-Davidson and the subsequent takeover. This amalgamation was to bring a new generation of mould-breaking muscle bikes, and facilitated the transition of Buell from a tiny producer of limited editions to a true manufacturer.

The first bike to roll out as part of the new era was the S2 Thunderbolt, a generously proportioned machine which could offer a measure of weather protection to the rider. Although sport-orientated, it was really what the Japanese describe as a sports tourer. There was even a police version of this and it was used by several departments. Then came the 'proper' touring version of this, the S2T, complete with panniers and leg shields.

From this evolved the S1 Lightning, a minimalist bike devoid of all but the barest of necessities and looking very different from anything else on the market. A cheaper, entry-level machine was the M2 Cyclone, with conventional forks, as opposed to the inverted ones, and a lower state of tune.

Further variations on these models has since produced the S3 Thunderbolt, the White Lightning and Lightning Strike and the X1, but they are still based on the same principles and basic themes. All are selling well and attracting a growing band of fans that might not have otherwise considered a bike with a Harley engine.

But all these bikes were still at the enthusiast end of the market: what Harley-Davidson required was a genuine entry-level machine for first-time riders who,

Buell
Blast!

OPPOSITE: Blast! The first small Buell ever – in fact the first small civilian Harley for over 20 years.

ABOVE: Erik Buell, left, being interviewed by *Big Twin* magazine.

having found their motorcycling feet, would go on to buy bigger, more expensive Harleys. Even the Sportster 883 was too large and intimidating for some, but the answer came in the form of the single-cylinder Blast! (the exclamation mark is part of the name) which was unveiled in 1999.

It utilized a new 30-ci (492-cc) single-cylinder motor, though it owed something to the existing Sportster V-twin, that was mounted in a steel backbone frame which rolled along on 16-in (406-mm) wheels to keep the seat height low. However, typical Buell design details could still be seen, such as the silencer under the motor and the large front disc brake, which add confirmation to the name emblazoned on the tank. It is a Buell, but not as people have come to know it.

Buell was quick to acknowledge that the tooling and processes used on the V-twin motor were used in the production of the single, along with a few components like the gearbox. However, it was also stressed that it was much more than an existing V-twin powerplant, chopped in half. The overall philosophy was simplicity and ease of riding, seen as most important in attracting the learner rider. The small stature of the bike, and the total lack of complication to enable the rider to concentrate on learning to ride, were also paramount in the execution of the design.

All this shows that Buell is not content to stand still and will continue to move forward, providing Harley with what it wants, and is an innovative arm that will try out new markets. This leaves Harley in a safe position and precludes it from compromising its important heritage.

Buell's Philosophy

Erik Buell has always been acknowledged as a deep thinker and a person who can recognize a problem and formulate its solution. There are two words, however, which encapsulate his philosophy of engineering: innovation and simplicity. When one looks at both the V-twins and the single, one can see that these principles are uppermost in the design and manufacturing of the Buell machines.

Buell also believes that the objective is to 'create motorcycles for enthusiasts who believe the fun of the actual riding experience is every bit as important as the destination itself'. Every Buell motorcycle appears to have been designed, engineered and manufactured with this in mind.

Buell insists that some bikes are becoming over-complicated and designed with top-speed as the ultimate goal. This, he argues, is only relevant on the race track or on unrestricted autobahns: for most riders, speed is irrelevant. So he provides a bike which can be ridden and enjoyed in a real-world scenario. Likewise, the style and form is designed with function in mind rather than with a view to pleasing design committees. This has led to the bikes' very distinctive style, one that is unmatched by other manufacturers.

Buell
AMERICAN MOTORCYCLES

OPPOSITE: Buells can be as much about looks as performance. This one has many aftermarket parts.

ABOVE: Buells and Harleys don't appear to be on speaking terms!

ABOVE RIGHT: Thunderbolt has a bigger seat and more bodywork than the Lightning.

RIGHT: Buells are more slimline, far less heavy machines than Harleys.

Apart from this, all the bikes have been built bearing in mind Erik Buell's three fundamental design concepts. These are: Mass Centralization, Frame Rigidity and Low Unsprung Weight.

Mass centralization is the practice of concentrating the greatest amount of mass as close to the centre of the motorcycle as possible. By positioning the engine, exhaust and suspension components of the motorcycle near the centre of the bike, frame stress is reduced and there is a lower polar moment of inertia. The horizontal rear shock-absorber not only helps to keep the mass of the bike centred, but also reduces weight.

Frame rigidity results in a machine that offers the rider greater control. To this end, the light 29-lb (13-kg), but immensely strong chrome-moly geodesic perimeter-frame design has many triangulated sections for maximum frame stiffness.

Minimizing the weight of those components not supported by the motorcycle's suspension springs is also critical to control and the feel of the ride. Low unsprung weight allows the tyres to maintain contact with the road, irrespective of the surface conditions. The six-piston, single-disc front brake on all Buell models is one way in which unsprung weight has been reduced, saving around 4–6lb (2–3kg) compared with conventional twin disc set-ups. Whilst saving weight, the 13.38-in (340-mm) disc and six-piston caliper also offers superb braking power.

Maybe it is the pursuit of these goals that have given Buells their quite unique look. One of the most distinctive features is the large-volume Helmholtz Resonator

HARLEY DAVIDSON

OPPOSITE: Lightning: Buell has become the definitive muscle bike.

RIGHT: This Victory cruiser has inspired Erik Buell to make best use of a big torquey V-twin engine, but in a sporting chassis.

airbox, which has been the subject of much criticim and has been seen as spoiling the bike's looks as well as restricting leg room. But there is a reason for its size. It is not an empty box: inside, tuned ducts and chambers allow free air-flow for instant throttle response and peak performance, while cutting down induction noise to meet European noise regulations. The X1 features the same large-volume airbox, but the design is now integrated within the form of the fuel tank cover for improved styling and as an answer to the styling critics.

Buell's Thunderstorm V-twin engine is built by Harley-Davidson, of course, but it is not the same motor you will find in the Sportster. There is more torque and power compared to the stock item, obtained by using different cams, lighter flywheels and high-compression cylinder heads with bigger valves and revised ports. The motor is fuelled by Buell's Dynamic Digital Fuel Injection (DDFI) system. Its closed loop design monitors conditions inside and outside the engine, constantly checking fuel, air mixture, ambient air temperature and humidity, engine temperature, barometric pressure and exhaust output to ensure optimum performance.

The stainless-steel exhaust pipes that offer a tortuous route into the tuned large-volume muffler, running under the bike, have often been the subject of comment, like the air box. The reason for the design is that it enables exhaust gases to flow more freely. Likewise, the large muffler has internals that are scientifically tuned to cancel out problem noise frequencies and allow the bike to pass noise tests. The patented Uniplanar engine vibration isolation system uses rubber mounts to isolate vibration from the engine, to which the swing-arm assembly is bolted. Rose-jointed tie-rods restrict the movement allowed by the rubber mounts to a vertical plane to ensure constant wheel alignment.

Allied to this, Buell, like Harley, uses a strong and durable toothed belt for the final drive. The belt is clean, quiet and smooth in operation, with low maintenance requirements, and its lightness contributes to the Buell's low unsprung weight.

As one might expect from his background, Erik Buell still uses the race track to test and validate technological developments and to improve the production motorcycles. They are now the important issue, but it is well known that, 'racing improves the breed', and allows bikes to be pushed hard and tested to their limits. Buells compete in the AMA Pro Thunder Series, and the experience gained is fed directly back to street-bike development.

Riding the Brand

Riding a Buell is different from riding any other Harley. While the expression, 'laid-back', may be appropriate to most of the Harley-Davidson range, it cannot be applied to any of the Buells and that includes the Blast, civilized though it may be. To give an idea of what Buell has created, here are a few examples.

S1 Lightning

The S1 Lightning adheres to the Buell philosophy in every respect. It is a sports bike, but an unconventional one. Visually the bike is different: it is devoid of the acres

OPPOSITE: Lightning at speed: that large black box beneath the motor is a silencer.

RIGHT: Buells have proved their worth on the track.

of plastic that one normally associates with a sports machine. Its looks are nonetheless striking, due to its abbreviated bodywork, large circular halogen headlight, and triangulated chrome-moly frame. But the critics are right: that large airbox on the offside, not to mention the huge silencer beneath the engine, are eyesores, disturbing the overall appearance. Buell trademarks they may be, but they are far from attractive.

At the time of its introduction, the first S1 was the lightest bike ever produced by Buell, weighing in at 425lb (193kg).

The mass is centralized and as low as possible to keep the centre of gravity low, so aiding stability and responsiveness according to Buell's basic design concepts. Allied to this is a short and stubby 55-inch (140-cm) wheelbase, which adds to its desirably responsive handling characteristics.

Right in the centre of the bike is the Harley 74-ci (1203-cc) engine, but in a very different state of tune to the one that powers the Thunderbolt. Buell utilized the skill of Harley engine tuner Gary Stippich to provide an engine that, as strange as it may seem, began life in the same factory. Most of the work was done on the top end of the engine, although 3lb (1.4kg) of metal was shaved from the flywheels.

Completely new cylinder heads were cast, and these now combine the combustion chamber shape of the smaller 54-ci (883-cc) motor, with the larger valves

91
SUPERMONO
buell
AMERICAN MOTORCYCLES
91

of the bigger engine. While they were at it, the compression ratio was also raised from 9:1 to 10:1. Some parts were replaced by Screamin' Eagle performance components, such as a cam with a longer lift duration.

As one might expect while the work was in progress, the inlet tracts were modified and the carb re-jetted to cope with change in demand. This also necessitated alterations either side to deal with changes in volumetric efficiency. Independent specialists were brought in to redesign the airbox and exhaust systems so that the bike would meet all the relevant noise and emission tests, not only in the US, but also in other world markets. These changes boosted power to a claimed 91bhp, being 15 more than a Thunderbolt engine and a massive 50 per cent more than a standard Harley.

The modifications have not affected the rideability of the bike. It is all usable power, allowing the S1 to pull all the way from tickover to the red line, with no fluffing or flat spots: there is raw power all the way through. Open the twistgrip and the large airbox gulps air in as the exhaust note rises and the horizon approaches. Exit a bend or corner with an open road ahead, feed the power in hard and the bike gallops towards the next.

Reduce speed for the approaching corner (normally backing off the throttle is sufficient, due to the masses of engine-braking available), take an appropriate gear, and drop the bike in, with the knowledge that it will hold its line without complaint. The inverted multi-adjustable WP forks at the front seem to have no problems dealing with any road irregularities, and only the worst of these cause a little twitch from time to time. At the rear and below, the single shock in tension does a similarly reassuring job keeping the back wheel in check.

ABOVE: Buells handle well due to a stiff frame and relatively low weight.

OPPOSITE: Battle of the Twins racing has allowed Buells to race competitively against Ducatis and other twins.

The reworked motor can be spun hard towards the 6,300rpm red line with no worry and if this is done the bike will go as fast as its name suggests. one can feel the engine working, but the vibes are not intrusive. The five-speed box is always precise when shifting through the various well-chosen ratios, but changes cannot be hurried if they are not to result in a clashing of cogs. Certainly, there is also no noticeable transmission snatch from the Kevlar belt drive, or for that matter delay in responsiveness.

The chassis too has no flex or ever seems to get out of shape, even when the bike is pushed hard, which is what you would expect from a design by an ex-racer, while, overall, the handling is always predictable when the bike is being ridden through the twists and turns of meandering roads. At the other end of the scale, relaxed riding is possible, but not necessarily recommended. The heat getting through to the rider can be uncomfortable, especially in heavy or slow-moving traffic, though of course this was never intended to be Lightning's environment.

On the freeway, legal speeds are fine, but the upright riding position, due to the straight bars, becomes tiring once legal limits are exceeded. Head towards the 130-mph (209-km) top-speed and you will need to hang on grimly, which puts strain onto wrists, arms and shoulders. This is not the machine for 'green' riders: if you head for the red line, or cruise at high speeds, it is a real gas-guzzler. Painful at any speed is the seat, which is too sparse and minimalist, as is the leg-room for those on the tall side. Pillion riders also get a raw deal, and should not be contemplated for any length of journey, unless their rear end is on the small side and you are good friends!

Time spent cruising at steady speeds on an straight and open highway gives one time to appreciate the quality of the bike, the deep lustre of the paint on the petrol tank, cowling and mudguard glowing expensively in the sun. One begins to appreciate the neat alloy housing for the clocks and sensible well-made switchgear and controls that are very un-Harley. In fact, one will be aware of all the attention to detail that makes the bike what it is: a hybrid of mass-production and the hand-built. Operation of all the controls and operating systems, from the wet clutch through to the brakes, is smooth and gives plenty of feedback. The massive front brake is excellent, the six pistons gripping the 13.4-in (340-mm) disc with enough power to slow from whatever speed one likes and deal with whatever weight of rider may be on board at the time. At the back end, a single cast-iron disc takes control, gripped by a Brembo caliper, which is sufficient when considering the front's effectiveness.

Top quality components are the name of the game in all departments, like the 17-in (432-mm) Marchesini cast-alloy rims, shod with Dunlop Sportmax tyres. Both are well suited to one another and the bike in general, and allow the handling limits to be explored to the full.

Erik Buell said that it was his intention to create a bike for riders who believe that the actual riding experience is more important than arriving at the destination. The Lightning fulfills this promise: it goes like a bat out of hell, handles well and more importantly stops when required. It has its faults, of course, but more than anything else, it is the embodiment of the Buell philosophy: enjoyment.

ABOVE: WR-series Harley racers from a different era.

OPPOSITE: The 30.5-ci (500-cc) Buell Blast offers ease of riding and a low seat height for novices.

The Blast!

At the other end of the scale is the Blast. Criticized for its name, that suggests a level of performance and handling it clearly does not possess, it is nevertheless an important model in the range and demonstrates the level of proficiency within the company and the fact that it is now looking to diversification.

The signs are that the Blast is already selling in large numbers; no doubt some view it as a cheap option or as a ride-to-work machine. Others will want it for its light weight and compact dimensions, not to mention low running costs. And of course, it is a genuine American, which to many is an important factor. But the Blast's main target is the novice rider, and is designed to give them a taste of motorcycling before they move on to something bigger. Consequently, the small, compact, single-cylinder machine is uncomplicated and user-friendly in all departments.

The plastic bodywork is virtually indestructible and will resist damage if the bike is knocked, so reducing repair bills and making the learner more relaxed; it has a clean, almost maintenance-free belt drive, together with ease of use. In theory, all the novice has to do is acquire the on-road skills. To help remove anxiety regarding its perceived complexity, the Blast has an automatic choke, like a car, easy-to-use controls, minimal instrumentation and nothing more to do apart from filling the tank with fuel and putting air in the tyres. It is significant that it is not only private individuals that are buying Blasts, but also bike-training schools.

The Blast is surprisingly plush and comfortable, and despite its small size can be ridden by a 6-footer with no trouble at all. The bike's main frame is composed of a wide steel beam which forms a backbone and an attachment point to which the other frame parts are mounted. These consist of a box section and formed flat plates, rather than the round section tube we are more accustomed to seeing. The main section is also an integral oil reservoir to make use of the space and reduce weight. The engine is mounted using the Uniplanar system with which Buell riders are now familiar.

The 30-ci (492-cc) motor is relatively low-tech in some respects, having just two valves operated by hydraulic adjusters, which are self-regulating. Cooling is by air and the unit-construction powerplant has just five gears contained within its case. There are no balance shafts to smooth out vibration, unlike Japanese singles; instead, it relies on the unique engine-mounting system, which actually allows the factory to dial in a few vibrations to get riders accustomed to them, so that when they move up to a full Harley engine they will be unfazed by their presence.

An automatic choke on the 1.5-in (40-mm) Keihin carb makes starting a one-touch affair (no kick-start) to get the single spinning, and it soon settles down to an even if lumpy tickover. It is then a case of engaging the first gear, that has the inevitable clunk from rest, and moving out into the traffic to experience the full Blast sensation. The exhaust note has been tuned to make the bike sound right and in fairness it does sound pleasant as one pulls away from rest. Around town, or using the available power to go through a series of bends, it also sounds good, spot-on, in fact. However, at cruising speeds, the wind cancels this out and the exhaust is almost inaudible above the rush of the wind.

At low speeds, the built-in vibes are noticeable. However, like most American motors, once out on the open road, everything smoothes out and the bike will happily cruise at all legal speeds with enough in reserve to make it less tiring than when it is at full stretch. In traffic, it is nimble enough to allow gaps in traffic to be manoeuvred and the low-down punch of the engine means the gearbox does not have to be worked too hard: there is good throttle response as well.

Controls and layout are functional and easy to use. There are no frills – just the basic legal and safety requirements. The electronic speedometer with its LED odometer, is easy to read at a glance, while still giving sufficient information to the rider. Petrol consumption is quoted as an average of 70-mpg of mixed riding, which is quite useful considering that the tank capacity is only 2.8gal (12.7 litres), and means that there could be around 200 miles (322km) between fill-ups.

The bike copes as well with two up as it does with one and the small five-spoke 16-in (406-mm) wheels, shod with Dunlop tyres, allow the bike to track well through

Blast!
Buell
Blast!

sweeping, flowing bends. It is just as efficient on tight ones, due to the good overall balance and weight distribution. At the front, the 1.5-in (37-mm) conventional Showa telescopic forks also cope well, considering that there are no means of adjustment present. While this is not uncommon on the front, no adjustment on the rear suspension, not even pre-load, seems a strange omission. The idea, of course, is to reduce the complexity of riding the bike: it must be remembered that there are experienced riders who have no idea how to set up the suspension system for optimum handling and comfort. However, the Showa gas-charged shock seems up to the task and the bike is unlikely to get out of control or ran out of ground clearance. The traditionally large (for a Buell) single front disc brake has no trouble slowing the Blast's relatively light weight.

There are a host of factory extras, such as tank bags, panniers and tail racks to give the rider a chance to use the bike for touring or for carrying shopping. For those wishing to be different, there is a rather tacky decal kit available as a factory option to allow one to personalize the bike.

All in all, when looking at the Blast in relation to the market for which it is intended, there is little to criticise. Experienced riders may want more, but in reality, provided one is willing to work the bike hard, it provides an acceptable balance between function and fun.

Initial production has had to be stepped up to meet demand, so it looks as though this latest extension of the Buell philosophy has succeeded. Inextricably part of the Harley-Davidson story, it looks as though Buell will remain there for some time to come.

OPPOSITE: Harley-Davidson hopes the Blast will take it into a whole new market.

LEFT: Five-speed gearbox and single-cylinder engine have much in common with Harley components.

RIGHT: Blast is styled to look like the bigger Buells.

THE RACING GAME

It would be impossible to catalogue all of Harley-Davidson's racing exploits. Over the years, the bikes from Milwaukee have competed on board tracks, dirt tracks and tarmac. They've climbed cliffs, hills and mountains, raced on the road and off it, indoors and out and with short circuits and gruelling endurance runs, they've done the lot. It is all the more surprising when one learns that the four Founders, William Harley and the Davidson brothers, were no great fans of competition, seeing it as an expensive distraction from the main business of making and selling bikes for the road: in fact, it would take over ten years for them to be persuaded to officially enter a race. This was the 300-mile (500-km) event at Dodge City in 1914, and even then Harley failed to win! But from this inauspicious start, Harley-Davidson was to go on to be a major force in almost every aspect of motorcycle sport. This is so large a subject that it would fill a complete book, so we shall concentrate instead on circuit racing in all its forms.

In fact, the company had officially taken part in competition before that Dodge City race, but these were endurance and reliability runs, not races, the trick being not to finish first, but simply to finish, which with pioneer bikes on rudimentary roads, was an achievement in itself.

A board-track racer and FLXI 'flexy' sidecar.

Walter Davidson himself had entered and ridden in these events, notably the two-day endurance run to Kokomo in Indiana in 1907, and back to Chicago. He was one of only three riders to complete the 400-mile (650-km) event with a perfect score, which was no mean feat considering that roads were little more than horse trails in places. A year later he did the same on a belt-drive single-cylinder Model 4, from Catskill, New York to New York City, completing the run with a perfect score, which was good publicity for Harley-Davidson. Both events revealed the ruggedness and reliability of these early Harleys, something the company was keen to accentuate, even in those very early days. If it were to survive in competition, it needed to demonstrate the extreme reliability of its bikes, rather than their speed, and endurance runs did just that.

Racing, on the other hand, involved bikes being stripped and tuned, often over-tuned. They could blow up, crash or break down, giving the impression that they were unreliable. This, in other words, was the

Not all surviving board-track racers are museum pieces: these bikes are part of a demonstration race at Davenport, Iowa. None of the board tracks themselves have survived but they were relatively cheap to erect and allowed high-speed racing in almost every county. But the crashes could be horrific, and they were superseded by the dirt-track ovals which took precedence in America over European-style twisty tarmac circuits.

antithesis of the Founders' philosophy.

But this didn't mean that Harley-Davidson had no involvement in racing. There were some very good privateer riders around, keen to ride Harleys, so the solution was to quietly build race bikes for certain favoured riders, which Harley did, as early as 1910. If they subsequently won, all well and good. If they didn't, then their entry could be regarded as a purely private affair, and nothing to do with the factory!

Around this time, the 'Speed Bowls' or 'Thunderdromes' were also popular, with bikes being raced on 60-degree wooden slopes. Speeds of up to 90mph (145km/h) on rotting wooden boards were achieved by bikes with no brakes and cut-off exhaust pipes and often meant that riders were badly hurt, splinters being the main cause of injury. This led to the nickname 'Arena of Death' and initially, at least, Harley had no wish to be associated with them, especially after the 1912 catastrophe at Newark, when a bike spun into the crowd, killing spectators.

Arch-rival Indian, along with many other marques, had a different attitude to racing, and was never slow to capitalize on achievements in the sporting arena and extract the maximum media coverage available. It was also taking part in events outside America. When it took the hat-trick of wins at the 1911 TT races on the Isle of

ABOVE: Early singles raced the board tracks as well.

OPPOSITE: This is Steven Wright's restored eight-valve racer of 1915/16, which was Harley-Davidson's first serious stab at a purpose-built racing bike.

HARLEY-
DAVIDSON

Man, the publicity was immense and Indian bikes sold like there was no tomorrow. Harley-Davidson noted this and began to advertise its privateer wins; but to a certain extent it was still sitting on the fence, as was seen in a 1913 desert race at Phoenix, Arizona. The Harleys failed to do well, and the factory covered its shame by hastily pointing out that the entries were private, not factory-sanctioned!

In fact, that same year it issued a statement to the effect that 'race victories are just a needless desire to impress and something a serious manufacturer has no need to do'. But behind the scenes it had decided to adopt an official racing policy, though more out of commercial necessity than any genuine love of the sport.

RIGHT: The 1916 F-head motor wasn't as powerful as the eight-valve, but was often more reliable.

OPPOSITE: The F-head V-twin formed the backbone of Harley-Davidson's early racing effort, though special racers from other manufacturers soon persuaded it that something more exotic was required.

HARLEY-
DAVIDSON
10

LEFT: An eight-valve racer being fettled at a modern race meeting.

OPPOSITE
ABOVE LEFT: A 1916 board-track racer. The white tyres, wheels and frame are period items.

ABOVE RIGHT: A 1922 74-ci (1200-cc) JD board-track racer with exposed chain drive and extreme riding position.

BELOW LEFT: A 1920 61-ci (1000-cc) racer, as ridden by Dewey Sims.

BELOW RIGHT: An eight-valve board-track racer. Note the exposed pushrods for the four valves per cylinder and friction-type damper for the leading-link front forks.

The Ottaway Era

William 'Bill' Ottaway had been poached from Thor, another fledgling manufacturer in the US at this time, and one which was doing well in competition. Ottaway was largely responsible for this racing success and was given the task of setting up Harley-Davidson's race department. As a result, he was to be central to the company's early golden era of racing victories.

Harley-Davidson's reasons for this dramatic U-turn in policy may have occurred when the Founders realized that they had already achieved their initial goal, without compromising the principles on which the company had been based. The bikes were solid and reliable, capable of lasting the distance; now they needed to make sure they were the first to cross the finishing line, which would be Ottaway's job. Competition among manufacturers was growing more intense, with smaller players falling by the wayside: racing seemed to be just another important weapon in the growing sales war.

HARLEY-
DAVIDSON

HARLEY-
DAVIDSON

Bill Ottaway certainly came up with the goods in the form of an 8-valve V-twin (four valves per cylinder), based only loosely on the existing 61-ci (1000-cc) engine, which cost over $25,000 to develop. Part of these costs went to Englishman Harry Ricardo, the engine breathing specialist. He solved the intake and exhaust problems that were initially experienced with the engines, and when he had finished, they were putting out a very respectable 50bhp.

Well-heeled private racers could in theory buy one of these new 8-valvers from Ottaway's new racing department at Milwaukee, but at $1,500 each there weren't many takers. There was a rumour that the high price was to discourage privateers, enabling the best bikes to be kept for the new factory race team. For the first well-publicized race on dirt, which coincidentally was on 4 July, Harley entered five brand-new type-11K 8-valve machines, all personally prepared by Ottaway himself. It was not a fairytale debut, and there was strong competition from eight-valve Indians: the Harleys were not placed, although all finished the course.

The factory was not impressed, and expected better things considering the money it had invested in the new racing project. However, matters were to improve and by the end of that year it had claimed a championship win at Alabama. The next year, Harley was featuring a little more prominently in the results as its knowledge and expertise in the world of racing increased. In the road-racing series Otto Walker took a couple of prestigious wins and the opposition began to realize that Harley was now serious in its intentions towards racing.

When the 1916 racing season began, the famous Wrecking Crew was formed. The name was coined by journalists of the time, but was descriptive in describing the

ABOVE: A 1926 61ci (1000-cc) board-track racer V-twin. There was nothing to these early racers, which were little more than bicycles with overpowered engines bolted in.

RIGHT: Rear view of Steven Wright's eight-valve board-track racer.

OPPOSITE: A museum reconstruction of a board-track Motordrome scene, courtesy of the Otis Chandler Collection.

HARLEY-DAVIDSON
HARLEY-DAVIDSON
MILWAUKEE

HARLEY - DAVIDSON

HARLEY-DAVIDSON

OPPOSITE
LEFT: A single board-tracker: how simple can a racing engine be?

RIGHT ABOVE and BELOW: A 1922 74-ci (1200-cc) JD board-track racer. But a year before, Harley had withdrawn official support from the racing programme.

RIGHT: An original factory racer, the only known complete and original pre-1916 bike, with original race 'M' motor.

way Harley wrecked other teams' chances of success. Brought together by Ottaway, this hard core of riders changed over time, but some names remained in the picture and included Jim Davis, Ralph Hepburn, Fred Ludlow, Otto Walker, Maldwyn Jones, and Ray Weishaar.

Harley not only employed the best riders and kept them trained, it also provided the best back-up and logistical support. It is rumoured that Ottaway trained them with military-like precision, for pit stops and tactics, pit signals and so on. Whatever the truth, the results gave credence to this, with 15 National Championships coming its way in 1916. On the subject of tactics: the 8-valve machines were fast but still not as reliable as the production-based F-head bikes. So a favourite ploy was to enter teams consisting of both. In the early stages of a race, the 8-valvers would storm away to the front, daring other riders to keep up. When they and their rivals were sidelined, the slower F-heads were sitting near the back of the field, ready to take the lead.

At the end of 1916, top-level motorcycle sport in the US ceased after there was an agreement between all the major factories, who were gearing up for military contracts, and it would be three years before racing again began in earnest. The line-up of the Wrecking Crew had by now altered slightly, but their lust for victory had increased and one could see that they meant business straight away. New pit crew and extensive practice drills meant that a bike could be refuelled, given oil and a new rear tyre in 38 seconds, which is quite something even by today's standards and gave them a big advantage over the opposition.

One record after another began to fall to Harley and the team won the most important race of the year in Indiana, taking second and third place for good measure.

Meanwhile, the endurance runs had not been shelved, and Harley-Davidson dealer Hap Scherer set several new records on the new Sport Twin, which was designed more for reliability than outright speed. He rode it from New York to Chicago, covering the 1,012-mile (1630-km) route in 31 hours and 24 minutes. He also completed the famous 'Three Flags Run', from Canada to Mexico on a Harley.

HARLEY-
DAVIDSON
HARLEY-DAVIDSON
OF SUFFOLK
BARON
2

Incidentally, one of these bikes was also the first motorized vehicle of any type to climb 'Old Baldy' or Mount Antonio in California, whose 10,114-ft (3083-m) summit was accessed over a 7.5-mile (12-km) burro trail and demonstrated that the lightweight machine was as robust as it was light. Many a specialized vehicle had tried before and failed, so it was no mean achievement.

While 1919 may have been a good year in terms of results from the reformed race department, the following year was even better, with five national championships won that season. At the beginning of the year, Red Parkhurst competed at Daytona, setting several short-distance records, and using an 8-valve to take a combination record. Meanwhile, records were also being broken in the Dodge City race, re-run for the first time since 1916, and making up for Harley-Davidson's poor start at its debut event. In the same year, the origin of 'Hog' as the quintessential Harley nickname was established. Ray Weishaar had adopted a pig as his mascot at race meetings, and when he won the Marion in Indiana, setting a new record time in the process, he took it for a victory ride around the track. The pig, wearing a Harley vest, was named 'Hog'.

But despite success on the track, motorcycle sales were gradually decreasing and dealers and manufacturers alike felt that the bike had become more of a luxurious toy than a means of transportation. Cars were getting cheaper all the while, wresting the family transport role from the bike and sidecar. Walter Davidson was also increasingly frustrated that race success was failing to encourage more direct sales, and when the local police department at Dodge City still persisted in buying Indians, in spite of Harley's wins, it was the final straw. The Founders decided to pull out.

So although Harley entered the 1921 season with full works' support, it was to be its last for a few years, though the team itself did not know it at the time, neither did Bill Ottaway. Ironically, it had been its best season to date, the Wrecking Crew having won just about everything. By the end of the year Harley had won the 1, 5, 10, 25, 50, 100, 200 and 300-mile championships – in other words all of them – winning one race with an average speed of over 100mph (161km/h), the first time that this had ever been achieved.

Harley had spent $200,000 on racing that season and it is rumoured that certain riders had been paid as much as $20,000. But sales were failing to recoup this expenditure and the plug was pulled without the riders being aware of it until it was too late. (The cynical would describe this as a ploy to keep the riders committed until the end of the season.) The end was acrimonious, many of the Wrecking Crew having been left stranded in Arizona at that last meeting, with no money to get home. They were reduced to borrowing their train fares from a local Harley dealer and when they did get home, Walter Davidson refused to take their calls. It was

OPPOSITE: The Flexy was quite unlike modern racing sidecars in that it leaned around corners, just like the bike.

ABOVE: A board-track racing sculpture captures the essence of this close and dangerous sport.

LEFT: A 1927 21.35-ci (350-cc) OHV factory Peashooter. The little overhead-valve single was derived from a road model, and proved highly successful in Class A racing.

OPPOSITE: A later-model Peashooter with Art Deco-style tank finish. There were still no brakes, however.

Petrali & the Peashooters

a sad end to a glorious era.

Harley-Davidson may have pulled out, but its bikes were still being raced; even some of the Wrecking Crew continued on a private basis. There was also success overseas and Harley received some favourable publicity in 1923, when Englishman Freddie Dixon, riding a Harley, took the 61-ci (1,000-cc) World Championship at the Brooklands circuit in Surrey, England. In fact, two years earlier, a Harley had been the first-ever machine to lap the circuit at over 100mph (161km/h), ridden by Douglas Davidson (no relation).

With racing still so popular and with many more events taking place, the American Motorcycle Association (AMA) was formed in 1924 to organize the world of competition and act as an official regulatory body. One of its first problems was that the performance of the big V-twin racers was now outstripping the capabilities of their chassis. The number of fatalities and accidents was increasing, and the obvious solution was to slow race bikes down by reducing the engine size permitted. Harley-Davidson was ready to take advantage of this, having developed 21 and 31-ci (350 and 500-cc) single-cylinder OHV machines for the export market. Speedway was particularly popular in England and Australia and the bikes were designed with this in mind. Bill Ottaway, Arthur Constantine and Hank Syvertson developed the motors so that they were competitive from the start,

but they also formed the basis of one of the most famous racing Harleys ever, the short, cobby little Peashooter. This lightweight 350 single was to feature strongly in various race disciplines right through the 1920s.

In 1925 Harley made something of a comeback to racing when Joe Petrali beat all the established top-name riders at Pennsylvania on a twin-cam Harley. Relatively unknown until then, he had been riding Indian singles until the age of 21, mainly on the board tracks that still proliferated around the USA. He turned up at the meeting without a bike, but managed to talk his way onto a works Harley, replacing the injured Ralph Hepburn. He began the race as an unknown and ended it as a legend, having averaged 100mph (161km/h) for the 100-mile race and beaten all the established stars along the way. This was the start of a long racing career in which he was to maintain a virtual monopoly of the chequered flag for the next ten years.

To give an idea of Joe's record: he won the National Hill Climb Championship every year from 1932 to 1937, and speedway races in record times. From 1933 to 1936 his wins included the 200-mile (322-km) Expert Championship, the miniature TT championship, the 200-mile Speedway Championship and all 13 Dirt Track Championships. He also set four AMA records along the way. In 1935, for instance, he took first place in every single National Dirt Track race of the season. At one event, he managed five wins in a day at a New York meeting. Not content with actual track racing, in 1937 he took a streamlined Knucklehead Harley to a world record speed of 136.183mph (219.15km/h) at Daytona Beach.

One might imagine that a rider of this calibre would have received handsome remuneration from Harley for his trouble, and given every conceivable help there was; but this was not the case and would eventually lead to a parting of the ways. Even that new Knucklehead record at Daytona failed to net him a bonus, because he had failed to reach 150mph (241km/h)!

Meanwhile, with the country in the grip of the Depression, and the manufacturers more inclined to concentrate on survival, racing was in the doldrums. In an effort to give it a boost, the AMA introduced Class C racing in 1934. This was designed to be cheaper and more accessible, attracting more amateur riders without destroying the existing Class A and B professional factory-backed events. The bikes in Class A (21-ci/350-cc ohv singles and 45-ci/750-cc side-valve twins) and B (45-ci twins or 80-ci/1300-cc fours) were pure racers, but in Class C, the bikes (limited to 30.5-ci/500-cc ohv or 45-ci side-valves) had to be production-based.

As it happened, both Harley and Indian both made machines ideally suited to the 750-cc end of the class, which was hardly surprising, as both were highly influential with the AMA, having rescued it from financial ruin. Because of this, the AMA was obviously not totally independent at that time, both companies having a commercial interest in the rules, something that would later change. But for the moment, the rules undoubtedly favoured home-produced bikes over the imported ohv singles.

The initial rules also stated that bikes had to be ridden to meetings, so Harley sold its WLDR fully equipped with all road-legal equipment such as lights. Moreover, the bikes had to be part of a production run of at least 25, to prevent the use of one-off specials, and even had to be offered for sale after winning a race for a maximum of $1,000. A sound marketing move, perhaps, but Harley ended up finishing behind Indian in most events. So in 1941, Harley marketed two bikes solely for racing, the WR and WRTT. The W indicated the 45 engine and the R racing and it was the sort of bike that the dealer and private owner would end up with after stripping down the WLDR.

This appeared at a time that the rules had slackened and because many of the bikes were now being towed on trailers to meetings. These new models already came stripped of heavyweight road-legal equipment. Tuning was becoming *de rigueur* and Harley was listing many

The WR racer is still ridden in anger, on the tarmac circuit at Daytona (below) and on dirt track (opposite).

6P
6P

ABOVE: WR racers still relied on hand gearchange. There is a Hummer racer in the background.

OPPOSITE: The WR was a bitter rival of the Indian Scout.

tuning items in its parts catalogue. However, the rules had the effect of creating a level playing field and if a rider did not win, it was now down to him rather than the bike. By 1938 the AMA had actually suspended the A-class races as there was now little interest, due to the success of the C-class having captured people's imagination.

Of course, racing was also taking place on the dirt ovals and the sight of 130-mph (210-km/h) V-twins broadside with no brakes made the European Speedway scene look rather tame. It was at these events that riders learned to control 45-ci (750-cc) machines on the slide and taught riders how to totally master the bikes so that, on returning to tarmac, they cleaned up there as well. It was this dirt-based apprenticeship that made so many American riders adept at high-speed slides, and was one reason why they went on to dominate world-class racing in the 1970s and '80s.

1
46x
Indian
Indian
HARLEY DAVIDSON

The Enthusiast
A MAGAZINE FOR MOTORCYCLISTS
MARCH
1940

HARLEY
DAVIDSON
STEVE COX
213
HARLEY-DAVIDSON

OPPOSITE
LEFT: March 1940 was a time of war in Europe, while racing in the USA continued, as this cover of Harley-Davidson's *The Enthusiast* magazine shows. Races at Daytona really were run on the beach.

ABOVE RIGHT: Steve Coe, a well known WR racer, with his son at Daytona.

BELOW RIGHT: WR racers aren't yet valuable museum pieces, or at least, not enough to prevent amateur racers like Steve Coe using them for their intended purpose.

RIGHT: A WR racer in action at Daytona. The classic races are an integral part of Daytona Bike Week.

LEFT: These two late-model WRs, both owned by Max Middlebosch in Holland, appear to be fitted out for the road, with moderately high bars and a silencer.

OPPOSITE
ABOVE LEFT and RIGHT: KR dirt trackers like the WR can still be seen racing. On the left is Rusty Lowry.

BELOW LEFT: The KR marked a departure from the WR, with unit-construction four-speed foot-change gearbox and hand clutch.

BELOW RIGHT: The engine, however, remained a side-valve, seen here in twin carburettor guise.

Post-War Challenges

In 1941, shortly after Harley had launched the WR production-race machines, all motor sport ceased due to the Second World War. It would not be until 1948 that things got back to normal, when riders on Harley-Davidsons took 19 of the 23 American Championships. They were sweet victories, as the year before Harley had been trounced at Daytona by an Indian ridden by Johnny Spiegelhoff.

But the real challenge came not from Indian, but from imports. Politically-motivated, many of the import tariffs had been dropped in the US, and there was a flood of British bikes in particular taking advantage of the fact and that overall they were considerably cheaper than Harleys in terms of cost. While the racing rules did not favour them, their nimble handling and foot gearchanges meant they were still able to win races. It also encouraged the British factories to speed up development and they were soon challenging Harley on the track as well as in the showrooms. The rule bias meant that the engineers worked harder to extract the necessary horsepower and top speed from the smaller engines, so in effect their progress was accelerated by the restrictive rule book.

By 1952, the ageing WRTT was struggling and was now well off-pace. Fortunately for Harley-Davidson, its replacement, the KRTT, was ready and

94
94

33

98
98

waiting to become the most successful racing Harley ever. This side-valve 45ci (750cc) was the most radical innovation ever made to the famous V- twin, and gave the company its first-ever Daytona win in 1953. Ridden by Paul Goldsmith, the 37-bhp KR managed to stay ahead of the opposition to take the flag. This was really the first full season for the race bike, having been launched late in 1952, and with the inevitable teething problems that needed to be resolved.

The KR had taken many of its new features from the British bikes that were now doing so well. There was swinging-arm rear suspension, controlled by twin shock-absorbers, and a four-speed transmission with foot change was the order of the day, with a hand-operated clutch immersed in oil. The telescopics and newly-sprung rear ends may have made for a good ride on the street, but the track revealed faults and they were difficult to ride. Harley eventually resolved the problems and factory rider Joe Leonard ended the 1954 season as the winner of the AMA No 1 plate.

LEFT: 24 hours, 1,366 miles (2200km), and an average 57mph (92km/h) – enough for a WR45 to win this marathon.

ABOVE: With a swinging-arm rear frame, the KR was used in road racing as well.

OPPOSITE: But as with the WR, its real forte lay in dirt-track racing – the heart of American competition riding. This is the 1/2-mile circuit at Davenport, Iowa.

DPC Racing
20
DPC
Racing
HOGWASH
20

884
302
Harley Davidson
784

OPPOSITE: Low bars, rear-set footrests, a swinging-arm frame and steering damper mark this out as a road-racing KR. A large side-mounted air cleaner dominates the engine.

ABOVE: The neat side-valve KR motor was competitive against the smaller ohv British bikes.

RIGHT: A later-model dirt-track KR, with swinging arm frame.

There was even talk in the press of a brand-new Wrecking Crew; but success was down to tuners like Tom Sifton, who had made it all possible, rather than the factory providing a winner from scratch. Once a Harley dealer involved in competition, Tom Sifton relinquished his business to concentrate on his first love, racing. His attempts at boosting power on the KRTT were directed into porting cylinders, improving cam profiles and fitting twin-valve springs. Moreover, there were a host of other improvements, some of which went on to be adopted by Harley-Davidson itself. It could be said that it was due to Sifton's expertise that Leonard won the title in 1954, with eight wins in the 18-race series.

The bike was sold to his friend, Leonard Andres, at the end of the season, who sent his son Brad out on it in 1955. Brad took the flag first in the season-opening 200-miler (322-km) at Daytona, winning at a record average speed of 94.57mph (152.2km/h). This was to be the start of a season that saw him win just about everything going and he ended it as number one, receiving the No 1 plate for the next season. This was the first time that a rookie had won the coveted plate at his first attempt. The icing on the cake was that Harley riders also won second and third spots in the championship.

With Harley back on top, the AMA changed the rules to allow a higher 9:1 compression ratio (the previous limit had been 6.5:1). This was actually a great boost for the British, enabling them to make use of their tuneable engines. But in spite of this, Joe Leonard reclaimed his title from Andres, who finished second. Harleys took a win in every one of the seven title rounds that year, as well as a host of chequered flags everywhere else they competed.

Leonard again took the plate in 1957, but he was soon to be beaten in the record stakes by Carroll Resweber, who finished fourth in that year's rankings. However, Leonard had also claimed his first Daytona 200 win and it was to be the year that saw the appointment of race manager Dick (OB) O'Brien at Harley-Davidson, a man whose name was to become synonymous with the Harley racing effort until well into the 1980s.

Resweber moved up to take the No 1 plate in 1958, which he managed to retain for four years in a row, all taken on Harley-Davidsons. He may have continued but for a racing accident that badly smashed his leg. During this dominance, however, other Harley riders were emerging to take the limelight, among them Bart Markel and Roger Reiman, both destined to become AMA champions in their own right.

Aermacchis Onto the Scene

In April 1960, news came that Harley-Davidson had taken over Italian bike maker Aermacchi. Its 15 and 21-ci (250 and 350-cc) horizontal overhead-valve singles had achieved a high degree of success bearing Harley badges, and they were also to do well in US competition. But in larger capacity racing, Harley was about to face a fresh challenge from Britain. Resweber took his last AMA plate for Harley in 1961, leaving the way open to Markel to retain it for Harley the following year. But the opening Daytona race was not good news for the Harley factory, as it failed to add another 200-mile (322-km) victory to the tally when Don Burnett gave the British Triumph factory a win, riding a 31-ci (500-cc) twin-cylinder bike in the prestigious race.

A year later, another two British factories would disrupt Harley history when BSA/Matchless-riding American legend, Dick Mann, took the AMA plate from Harley as well. He took the title by just one point from George Roeder on a Harley, after a dispute concerning his Matchless G50.

The same year, Harley took the inaugural 250 AMA National Championship title, which took place at Illinois and Bart Markel won it on an Aermacchi-Harley. The Sprint model sported a dolphin fairing and was capable

OPPOSITE: A 54-ci (883-cc) racer competing in the French Cup at the Montlhéry circuit, Paris.

Aermacchis took to the dirt track as well. ABOVE: A 21-ci (350-cc) Sprint and a 15-ci (250-cc) ABOVE RIGHT.

of 116mph (187km/h), which was not too bad when compared with the more powerful KR, which speed-tested the same year at 142mph (229km/h. In fact, the little Aermacchis were to do well both on the track and in short-track events until the Japanese twins took a hold from the mid-1960s. From then on they were unstoppable in the smaller classes although it would still be some years before they managed to take over in the larger capacity races as well.

In 1964 Harley once again took the overall championship due to the consistent skill of Roger Reiman. Like Dick Mann the year before, it was the consistent placing rather than the number of outright victories which brought success. However, one win he did gain was Daytona and although he repeated the victory a year later, it was Bart Markel who eventually stole the AMA plate. Markel was to do the same again in 1966, though the KR was now starting to struggle against the British bikes, which would capture the championship for the next two years.

It should be mentioned that getting a side-valve KR to reach 145mph (230km/h) was something of an achievement and the bike was still competitive in the US at least. It was still capable of winning Daytona as late as 1969, when Cal Rayborn gave Harley its last-ever win in the 200-mile event and which ended a 12-race score from 1955. The only bike to briefly interrupt that winning streak was a Triumph.

Outside the USA, Harley racing success was limited, though Californian Lance Weil did achieve some impressive results in England, racing an experimental 54-ci (883-cc) machine that won quite a few of the 61-ci (1000-cc) races he entered. Information gained from this foray enabled Dick O'Brien and his team

SHARK
SEGURA
ASSURANCE
moto verte
69
moto revue
DUNLOP

ABOVE: The 883 racing class has allowed lightly modified Sportster 883s to race as well.

OPPOSITE: A 54ci (883cc) at Montlhéry. Pipes, low-set handlebars and rear-set footrests are among the modifications.

to extract even more power from the ageing engine, enough to enable Rayborn to win his race at Daytona with an average speed of over 100mph (161km/h). In doing so, he also lapped all but the second- and third- place finishers.

In 1969 only eight bikes were built, and these remained with the official factory team. Featuring Italian Ceriani forks and a massive Fontana four-leading-shoe front brake it was sufficient to keep the Harley team up front, allowing Mert Lanwill to take the overall championship. But change was to come when the AMF takeover revealed that racing was not part of its great master plan: it was to be a lean time for the Harley-Davidson racing department.

It was, however, to spawn the legendary XR750, initially designed by Dutchman Pieter Zylstra. That first bike made its debut in February 1970 at the Houston Astrodome and was claimed to be an all-new bike. In reality it was little more than a revamped collection of parts, the chassis being a hybrid of KR parts and the motor coming from the XL Sportster. It retained the iron barrels and this was to be its initial downfall and meant that a revamped motor would appear two years later with aluminium components instead.

Those first bikes were fast enough, but they overheated and often failed to finish, earning the nickname 'Waffle Iron', although they fared better on the flat-track circuit. All the XR750s retired at Daytona in what proved to be Harley's worst-ever race season; its highest placed rider was Lanwill in sixth place in the overall championship. During the winter months, O'Brien and his team worked hard, but 1972 saw the bike beginning to suffer from gearbox breakages, although there were noticeable improvements to the engines.

The first sign that the XR750 was finally working out came at Easter in England when Cal Rayborn took a bike prepared by Walt Faulk to compete in the Anglo-American race series called the 'Transatlantic', and achieved success. He managed three wins out of six at three different race tracks and showed not only what a superb rider he was, but that the bike did indeed have potential, provided the problems could be cured. Rayborn also managed to give Harley some more positive publicity with an XLR-engined streamliner, by taking it to a land speed record of 265mph (426.5km/h) on the Bonneville Salt Flats.

Mark Brelesford was able to take the new alloy-engined XR to the AMA plate in 1972 and Rayborn returned to England and gave the legendary Jarno Saarinen, riding a works Yamaha, a run for his money until the Harley magneto failed. But this was to prove nothing more than a slight upturn of fortune and Harley entered the 1973 season with the factory as a whole suffering poor sales. On the dirt tracks, the thundering V-twins were losing out to the likes of Kenny Roberts on a Yamaha 650 twin. He was dominating the dirt scene and racing in general, taking the AMA plate before moving on to the world of GP racing.

ABOVE LEFT: Cal Rayborn's XR750 run by Team Obsolete.

ABOVE: American racing legend Cal Rayborn.

LEFT: Even the little 7.6-ci (125-cc) Hummer has been raced.

OPPOSITE

LEFT: Although the XR750 was designed as a dirt-track racer, it has taken to the circuits as well. This is George Roeder's bike.

RIGHT ABOVE and BELOW: Late-model Aermacchi Sprint racers with full fairings but still with drum, rather than disc, front brakes.

George Roeder

MOTUL
906

65

Daytona was also to pave the way to a poor season on the black stuff. Brelesford suffered an accident on lap 11, in which his bike was engulfed with flames after a crash. He unfortunately broke both legs and a hand, which was his racing demise. In the same race, Rayborn also crashed after his engine seized, resulting in a broken collarbone. Being a hard man, he was still able to captain the American squad in the Transatlantic shortly after, but he returned too soon and was not able to give of his best. He was also frustrated by the lack of Harley performance against the Japanese bikes and left at the end of the season.

This was not only unfortunate for Harley, but was also a loss to the world of racing, as the 33-year-old was killed when riding a Suzuki the following year. Many people believe that Rayborn, although never crowned champion, was one of the best riders the US had ever produced and consider him on a par with that other Harley legend, Joe Petrali.

RIGHT: Built for Daytona: George Roeder's XR on the infield.

FAR RIGHT: The XR750 in its element in a classic dirt-track.event. Scott Parker leads the field.

1
31
91
56

LEFT: Preparation is all: Chuck Davis' XR being prepared for an outing at Daytona.

OPPOSITE
ABOVE LEFT: A dirt-track XR, very different from the road racer, though it now also uses a swinging-arm frame.

ABOVE RIGHT: The XR750 engine is now a distant relative of the road-going V-twin, with a short stroke and twin carburettors.

BELOW RIGHT: There are high-level pipes on all XRs, whatever the discipline.

Revival

But if the early 1970s was a bad time for Harley racing, 1974 was to be its finest hour, albeit that the bikes were basically Aermacchi racers now wearing the AMF/Harley-Davidson logo on their tanks. Based at Varese, Italy, the team was led by Walter Villa, one of two brilliant rider-designer brothers who had evolved a brand-new water-cooled two-stroke twin. Villa began the season by taking the lap record and winning his home Italian GP at Monza. This was the start of numerous wins that saw him crowned 250 World Champion at the end of the season, which was a feat he repeated in 1975, giving Harley-Davidson its second World Championship in a row. It was also looking good for 1976 with a 31-ci (500-cc) having been produced, although it failed to live up to expectations and was given to the Bimota squad to race, eventually coming to nothing.

Even if the engine did not live up to expectations, Villa did, and took the 250-cc (15-ci) title for a third year on the trot with a victory at the Czechoslovakian GP in August of that year. A week later he added

the 350-cc (21-ci) World Championship at the West German GP, giving him and Harley a world double and their last win in the world racing arena. The bikes were at that time unbeatable, but the factory could not afford to race the following year and it was a privately funded outfit that ran in the championships. Villa still managed a third place in the 250-cc championship, with team-mate Franco Uncini taking second, but came nowhere in the 350-cc class at all. The following year the Varese plant was declared bankrupt and was sold on to Cagiva.

Back in the US, things were not going as well and the name Harley-Davidson was gradually slipping from rostrum places on tarmac. On the dirt ovals, things were better and the big V-twins could still be seen taking the top places and entertaining the crowds.

It was here that they managed to give their riders enough points to take the No 1 plate, among them Jay 'Springer' Springsteen, who took it for several years in succession. He was fast becoming another Harley legend and a name that would be associated with the factory into the next century.

At the beginning of the 1980s it looked as though Harley and racing would be relegated to just words in the history books, unless it was classic events, where the old war-horses still showed their mettle. Then came the 'Battle of the Twins' (BOTT) class for twin-cylinder machines, which proved the saviour of Harley's racing heritage.

The AMA may not have liked it, with a full racing programme already established at most race meetings, but when the inaugural event took place at Daytona in 1981, 88 riders turned up to race. The names on the bikes they rode

read like a list of the world's motorcycle manufacturers. Four different classes were run, the displacement limit was set at 61ci (1000cc), and an eight-race series began.

Harley may have been there with a half-hearted effort based on the XLCR Café Racer, but Ducati took the first honours. After the 1982 season, when the series was well-established, Harley then decided to have a serious crack at winning, with 'Lucifer's Hammer'. Built by Dick O'Brien, it was to feature in the history books along with other great race bikes. Ridden by Jay Springsteen, the Hammer stole the win in the BOTT race in 1983, winning by 24 seconds from the following machine. This was ten years after the last factory bike appeared in the US, ridden by Gary Scott at Laguna; Harley was now back with a vengeance.

Basically, the Hammer was constructed from a road-race chassis taken from the early 1970s and fitted with the XR1000 engine taken from Harley's then street bike on sale to the public. The bike achieved 104bhp and was to go on and win the BOTT titles in the Grand Prix category for the 1984–86 seasons. However, Springsteen was not in the hot seat for, although he rode it a few more time in 1983, his contract was with the dirt-track events, which always took precedence. Instead, Gene Church took the ride and the championship and the bike went on to be sponsored by the Harley Owners Group (HOG).

The bike was actually run by dealer Don Tilley, who managed to keep the bike competitive until 1989 when it was replaced by a Buell-based machine. Although having a relatively short career, the Hammer had proved its worth and paved the way for Harley to make a comeback to full-time racing.

In 1998 Harley itself launched a one-make racing series for 883 Sportsters. Run on a semi-professional basis, the slightly modified street bike-series quickly attracted the public's attention. Strictly policed, the racing was tight and close as a result. That inaugural year's championship was actually won by a British rider by the name of Nigel Gale who had won 'Rookie of the Year' in 1986 and went on to ride for Bartels, the Harley dealer who sponsored race bikes.

LEFT: With Cal Rayborn, Jay Springsteen is one of the most accomplished XR750 riders of all time. 'Springer' became the youngest ever winner of the AMA National Championship in 1976 when he was 19, and if there was still any doubt, he quashed it by winning in 1977 and '78 as well, all three times as an official Harley rider. Despite being diagnosed with a rare type of diabetes, he was to accumulate 40 race wins on short track, TT, 1/2-mile and 1-mile dirt tracks, proving an able road user too.

OPPOSITE: Springsteen in full sliding flight on an XR750.

42
SPRINGSTEEN
9
BIG TWIN
42

Japanese to the Rescue!

By now, Harley had recognized the need to be back in racing and was once more in control of its own destiny, feeling that it was able to embark on a new project aimed purely at racing, the VR1000. This was to be a costly endeavour, not to mention an historic one, as for the first time ever the race bike would not be based on a standard production machine. It did make a road-legal copy of the race bike in limited numbers and it was on offer to the public for nearly $50,000, making it somewhat exclusive: a set of race parts was also included in the price for those who wanted to race them privately.

Harley announced that it would campaign the VR1000 in AMA Superbike and Grand Prix racing, but the first time the bike appeared was at Daytona in 1994, after five years' development. Once again it was a repetition of other bikes in Harley's racing past – no fairytale debut, but failure on its maiden run. The rider was Miguel Du Hamel, a top competitor and son of racing legend Yvon, who qualified nearly 8 seconds slower than the pole position Ducati. Cynics were beginning to

ABOVE: The VR1000 (seen here at Daytona) has so far failed to match the XR750's success on the tarmac circuits.

ABOVE RIGHT: Scott Zampach is one rider who has popularized a VR1000.

crow that in spite of Harley's massive resources, it could no longer build a race bike and compete with the world's best on the race track. When the bike lasted just 22 of the 57 race laps, due to the failure of a weld on the balance shaft, they seemed to have been proved right.

Two weeks later the bike failed to finish again, at Phoenix, Arizona, when a defective fuel line was responsible for starving the engine. It was to be a similar story the entire season, despite some creditable performances by the rider. Was it to be the early XR750 all over again?

The design team at that time consisted of a few well known names connected with the Harley brand: Jerry Branch, best known for his Harley cylinder-head work, Erik Buell, designer and bike builder, Dick O'Brien the former race boss, and last but not least, Don Tilley, the tuner of Lucifer's Hammer. The man in charge was Mark Millar, who actually got the whole project up to prototype stage.

The fairing was designed by none other than Willie G. Davidson, and it would later be his son who managed the race team in the late 1990s. Rousch Racing was used for getting the cylinder-head work done, thanks to its Nascar expertise, as a result of Harley's desire to keep the whole project as American as possible. Steve Scheibe did much of the initial work and he was eventually hired to take over the project.

The new engine retained the familiar Harley V-twin format, although it was moved to a 60-degree angle between the cylinders. The bore and stroke were 3.85 and 2.6in (98 and 66mm) and the cylinders nikasil-coated aluminium. The crank had two titanium plain bearing rods running on ball-bearing big ends, while the bike ran with a dry sump. Double overhead cams were used to operate four valves per cylinder and were chain driven, while a counterbalancer was driven off the crank to smooth things out. Water-cooling was used for the first time on a Harley motor, as was fuel injection.

Primary drive to the 'dry' clutch was by gears, the gearbox itself being five-speed. That first motor used at Daytona pumped out 140bhp at 10,400rpm with the rev limiter set at 11,000rpm. Weight-wise they were at a disadvantage, as well as having only test data to set the bike up, weighing in at 355lb (161kg), 20lb more than the minimum allowed weight under AMA rules.

While they could have used magnesium for crankcases, shaving 8lb (3.6kg) in one hit, this would have meant that all the road bikes produced would also have to use this expensive commodity, making them more expensive than they already were.

Eric Buell initially designed the frame, although his tubular design was later rejected along with a British design that was tried. Scheibe then used computer modelling to design Harley's own version, made by a company called Anodizing Incorporated which specialized in aluminium extrusion. The chassis finally ended up as a conventional aluminium beam with a massive head area for strength around the forks.

Despite attempting to remain totally US-based and initially using Penske suspension front and rear, Harley found that the Indy car team could not match Ohlins for handling. It therefore used the Dutch firm's 1.8-in (46-mm) inverted front forks. The brakes were taken from the car-racing world and Wilwood supplied the calipers, which featured pistons of different sizes for better load distribution across the pads.

Du Hamel was frustrated by the bike's failures and, unaccustomed to being off the pace, left at the end of the year, leaving Chris Carr (synonymous with Harley in dirt-track circles) and GP veteran, Doug Chandler, to get the bike onto the rostrum. Despite their considerable expertise in setting things up, progress was slow and it was not until 1998 that the bike began to be a serious contender.

With Canadian Pascal Picotte having moved into the hot seat, the Harley team

ABOVE: The VR1000 was a real departure for Harley-Davidson, with a water-cooled 60-degree V-twin and 11,000rpm rev limit. An aluminium beam frame and top-notch Ohlins suspension complete the specification.

OPPOSITE: Earl Small touches down a knee slider, trying hard on the VR.

EARL SMALL'S
HARLEY-DAVIDSON
MICHELIN
RACING
vanson
EBC

Scott Zampach riding a VR1000.

Chris Carr on a VR1000.

once again lost the opportunity to stand on the Daytona rostrum through poor pit-stop technique, something that would not have happened under Bill Ottaway's leadership. But it was not all bad news, for at that 1998 Daytona meeting, a Harley became the fastest-ever pushrod machine to lap the Daytona banking.

However, a Harley did win the Battle of the Twins F1 that year, with the all-black 'Daytona Weapon', which was making its sixth appearance in the Battle of the Twins F1. Constructed in 1992 by Japanese Harley guru Takehiko Shibazaki, it represented a totally different approach to the one the factory had been taking. Instead of a new high-tech engine, it featured the standard 45-degree V-twin, tuned and worked on by the man himself, who had previously built his own version of Lucifer's Hammer in the late 1980s. Shibazaki had himself given up racing to wield the spanner, and built the futuristic Weapon himself. He inserted his own tuned motor into a Japanese Over frame and using top components like Ohlins suspension and Brembo brakes to complement it, attempted to achieve reliability.

That first bike, like the factory VR, did not win on its initial outing, but the team kept coming back with improved versions. In 1997 another bike appeared called 'Golden Balls', so named because of its colour scheme. This was unfaired and the bike on which Jay Springsteen won the F2 race at Daytona.

In 1998 'Weapon 2' was brought along as well, complete with a new frame made by the British firm, Saxon. But this, like its forebears, did not manage to finish its first race, sitting it out to watch its older brother achieve that historic win. The rider was 'Mr Daytona', Scott Russell, taking his fifth win at the banked tri-oval circuit.

Harley was to go on to sign him to ride for the 1999 season, confident that the bike was now competitive enough to win in the hands of the right rider.

Before the race, however, a brawl in a Daytona bar prevented him from taking part and Pascal Picotte crashed out, leaving the team bereft of that much-coveted win. Harley was to be disappointed in 2000 as well, Russell failing to produce a win. So while the official Harley race team continues to persevere, it seems that its legions of loyal fans will have to wait before a new Wrecking Crew is formed and Harley is once again seen as a racing force to be feared.

OPPOSITE: Chris Carr on a VR at Daytona. At the time of writing, after five years of development and six years of racing, that coveted Daytona win has yet to happen.

RIGHT: Meanwhile, the Sportster 883 series provides an affordable means of racing for many amateur riders around the world.

Milestones: the Eight-Valve Racer

Designed in 1916 after Bill Ottaway had joined Harley from Thor, the eight-valve racers started life using already available basic components. Many engine parts came from the existing road-going F-head engine and the standard bore was kept, which gave the maximum 61ci (1000cc) allowable for race meetings.

However, the major change was to the cylinder heads, which gained two inlet and two exhaust ports per cylinder. A single camshaft located in the timing case operated the valves by long pushrods running up the outside of the barrels. The adoption of this design was for the normally accepted engine design principles of volumetric efficiency and better breathing. The design team used Harry Ricardo from England to make the four-valve system work to the best possible advantage after initially experiencing problems.

Fuel was supplied by one carburettor located between the V of the barrels and a Bosch Magneto provided ignition. There was still total-loss lubrication, although both a hand and automatic pump were provided to deal with racing conditions. (Englishman Freddie Dixon did run with a twin carb at Brooklands, but this was not a factory modification.) Three versions were produced with different rocker arms and cam formations. There was a one-shaft four-lobe camshaft, or one with two-lobe camshafts. Some engines had short exhaust pipes, others had none, relying on open ports to allow the burnt gases to escape very close to the rider's legs.

It would seem that no two engines were alike, due to the fact that only parts available at that time were used to build the engines. Power is thought to be around 50bhp, but varies due to the different compression ratios used to allow the bike to run on alcohol.

The motor was housed in a keystone frame, so called because it was a diamond shape dissected into fore-and-aft triangles. Plates at the bottom, which carried the bottom of the motor providing the main mounting points, joined the two frame down tubes, making for a rigid structure.

The rear of the frame was rigid with no suspension, while Harley leading-link forks were located in the front wheel with the occasional use of Flying Merkel forks. There were no brakes supplied, although a rear was listed as an optional extra for road races. Likewise there was no clutch, no

The legendary Jay Springsteen is pictured before the start of a dirt-track race (ABOVE) and at Daytona (OPPOSITE).

23
49
BELL
JAY
DUNLOP

OPPOSITE: An evocative sight: a string of XR750s chase one another towards the flag in the Del Mar races, California.

ABOVE: An Aermacchi Sprint, in road-race trim.

gearbox or starter mechanism, although some were listed for road-racing options, like the brakes. The bike was started by means of the compression release being dropped when sufficient speed had been obtained by a tow start, due to the high-compression ratios used.

The basic bikes were priced at $1,500 for the eight-valve and $1,400 for the four-valve single-cylinder version, which was the same bike with only the front cylinder fitted. This made it eligible for the professional classes run at the smaller tracks. Indian at that time priced its similar-specification bikes at $350 and $300! The number built and the specification is somewhat sketchy, estimates suggesting that somewhere between 20 and 60 bikes were made, the vast majority going overseas. Suffice to say that today they are rare – and still expensive.

The Peashooter

In 1925, when Indian brought out its budget-priced Prince with a 21-ci (350-cc) side-valve motor, Harley thought that it should respond. That response was in the form of two 350-cc side-valve machines giving 8bhp in both an A and B form costing $210 and $235 respectively. In smaller numbers, it also introduced the OHV AA with magneto ignition and the BA with a battery ignition system, both costing $250 and $275.

They were aimed primarily at the overseas markets, as the American buyer in the main still wanted the big V-twins. However, the sport of Speedway was growing fast abroad and the OHV AA-model quickly found favour with competitors. Tuned up and slimmed down to 215lb (97.5kg), the bike could easily top 80mph (129km/h). In the US it also found favour, and riders like Joe Petrali were almost unbeatable on it, often winning when competing with 30-ci (500-cc) machines. It also did well in hill-climb events as well as on the track.

Harley brought out a competition 'S'-version, which quickly earned the name Peashooter because of the characteristic dry popping sound made by its short exhaust pipe. This bike had a Sager front fork and 27-in (686-mm) wheels and cost $300. In 1926 it won 14 international titles and a year later all the US titles in the 350-cc class.

There were three basic versions: the prototype had an altered frame and special gas tank. The engine was fitted with a single port head and exhaust pipe, along with a Deluxe Schebler carburettor. This had no gearbox, just a countershaft with two different sprockets.There were flat-track versions with three-speed gearboxes and in 1927 there was another modified frame and a manual oil pump, a new carburettor and underslung handlebars.

One year on, the Peashooter was offered with a twin port head, two exhaust pipes and an Indianapolis carb. The engine also had a beefed-up con-rod for greater reliability. The export TT and road-racer had the three-speed box fitted along with the twin port head.

Another variant intended for A-class racing had a high compression ratio of 12:1 and was intended to be run on alcohol. There were no gears or brakes, the frame was very stiff with little suspension movement and the forks were set at a steep angle for quick, responsive turning. The bikes did well until the beginning of the 1930s when they began to lose out to machines fitted with the 30-ci (500-cc) British JAP engine that had been developed for Speedway use.

ABOVE: The classic foot-sliding technique demonstrated on an XR750 dirt-track racer.

OPPOSITE: An XR750, shown at the Sammy Miller Museum, Hampshire, England.

XR
750
HARLEY-DAVIDSON
38

The WR Racers

When Class C racing was introduced, Harley had to rely on the high-compression DL and DLD sports versions of the side-valve D-series on which riders could compete. In 1937 the new W-series succeeded it, with its own racing version, the WLDR, which developed 7bhp more than a standard WLD and cost $380. The WLDD that followed was a pure race machine formed by paying $109 for a conversion kit to transform the WLDR. The kit consisted of cylinder heads, cam, valve lifter and air intake.

However, the bikes always benefited from tuning and one well-known exponent of the model was Tom Sifton, a dealer from San Jose. His bikes, ridden by Sam Arena, took many race wins and Harley itself eventually incorporated many of his modifications into production models.

In 1942 the factory itself produced the WR and WRTT. The former was a standard production racer and was based on the type of bikes people were producing when they stripped the WLDR and did things made possible by a relaxation of the AMA rules. The WR weighed in at 300lb (136kg) and could run at a genuine 100mph (161km/h). The bike came with a vertically-mounted Edison-Splitdorf magneto, this later being replaced by a Wico unit. The motor itself had larger polished inlet ports and altered combustion chambers and stronger valve springs.

The engine came with boxes of spares and charts to allow it to be altered for the type of racing for which it was intended. Different gearing, wheel hubs and a choice of 18 or 19-in (457 or 483-mm) wheels were part of the package. The frame was strengthened for the rigours of competition and there were small or large gas tanks. In fact most of the bike was alterable making the rider and tuner's life that much easier.

There was also a WRTT special for road racing and hill climbing only and this had cut-down fenders along with clutch and brake pedals. This version had brakes compared to other disciplines where they were not allowed.

In essence, the WR race bikes brought Harley-Davidson's racing effort to a level playing field and allowed races to be decided on rider skill, rather than the amount of tuning they could afford or sponsorship they had received.

ABOVE: The WR was Harley's backbone racer for 15 years.

OPPOSITE: The RR250, on which Walter Villa won three World Championships for Harley-Davidson. This is the 1975 factory machine.

AMF Harley-Davidson
AMF Harley-Davidson
24

The KR

It was in 1952 that Harley-Davidson introduced the new K-model, a side-valve 45-ci (750-cc) V-twin which superseded the W-series. The KR version was available for the 1953 season and like most of the other Harley race machines it suffered initial teething problems. Despite this, it showed that it had the potential to beat the British bikes that were becoming so popular with the competition motorcyclists. In 1953, Paul Goldsmith used one to beat the opposition quite convincingly at Daytona in the 200-miler.

The KR had cylinder heads of alloy and the transmission was now incorporated within the same engine cases. Like the road-going K, it had a foot gearchange, four-speed gearbox and hand clutch, but made do with a small fuel tank from the little Hummer 125. However, it still used a single carburettor. Despite being underpowered for its engine size, and bearing in mind that it retained side valves when all the opposition had gone over to overhead valves a generation before, it had an incredibly successful 16-year racing career.

The frame of the KR was a smaller modified version of the K, made of superior grade steel. It featured lugs behind the seat, which enabled either a rigid or sprung rear end to be bolted in place, so that the same basic frame could be used for various competition disciplines: with a rigid rear end, it was ready for dirt-track racing, but bolt on the swinging arm and twin shocks and it was ready to road-race.

Drum brakes could be fitted to either the 18 or 19-inch wheels and in fact there were literally dozens of optional items available. Sprockets, handlebars, different sizes of gas tanks were all listed for the competition rider. So, with one bike and a little juggling of components, a rider could compete in short track, miles, half-miles, TT and road events, and in so doing be a national champion with just one bike! This policy came from the factory-inspired

ABOVE LEFT: A KR, set up for road-racing.

ABOVE RIGHT: A KR dirt-rack racer.

OPPOSITE LEFT: Hummers raced on dirt tracks too.

OPPOSITE RIGHT: In 1961, the KR was still Harley-Davidson's staple racer and was to continue to be for nearly two decades.

tuners like Tom Sifton, who worked hard to improve power. For this he used double-coil valve springs, sodium-cooled valves and chrome-plated piston rings, along with some general beefing up of components. The result was the bike that allowed Brad Andres to take his No 1 plate while still a rookie in 1955.

While the engines initially gave around 38bhp, they ended up with 64bhp in 1968 at the end of their lifespan. The 1968–69 machines, under the direction of race manager Dick O'Brien, were reworked and developed with a low-slung 'Lowboy' frame and a wind tunnel-tested fairing. The motor had a Tillotson dual-throat carb, hotter cams and altered timing in an effort to keep the other racing teams at bay.

It worked, and it was on one of these KRs that Roger Reiman achieved his record-breaking Daytona 200 result. In 1968, only eight works KRs were built, using Ceriani forks, Fontana drum-brakes at the front and a disc at the rear. But then a change in AMA rules took away the advantage of a big side-valver like the KR; now there was no limit on valve operation or the number of cylinders, and the KR, well developed as it was, stood little chance against three-cylinder Triumphs and Honda fours.

But on the dirt track they continued to be competitive for some years after, underlining how different this type of American sport was from the European.

The XR750

By the time the British factories had managed to get the AMA rule book altered so that they at least stood a fighting chance, the writing was on the wall for the KR, which had passed its sell-by date. It was surprising, therefore, that Harley was caught napping with nothing on the horizon to take the KR's place.

Race-team manager Dick O'Brien did what he could and took the iron engine from the club-race version of the Sportster; he reduced the stroke to make the displacement limit that still stood at 45ci (750cc) and put it into a KR frame. This makeshift was given the designation XR750, or the racing version of the X-series engine.

Once again there was an imperfect beginning: the early XRs were overweight, under-powered and somewhat fragile. If the engine did not blow it would tend to break down. Harley had complied with the rules and built 200 to be eligible to race. Unfortunately, word got out and sales were poor, with half the bikes being used as spares for the rest that were being raced. It was therefore no great surprise when BSA and Triumph took the championship in 1970 and '71.

It should be remembered that this was taking place at the same time as the AMF takeover and Harley's new bosses were not fans of racing. Undaunted, O'Brien continued to work away, and in 1972 he launched the new version with alloy heads bolted through to the crankcases, improved combustion chambers and a better one-piece crankshaft.

The engine was over-square (the bore was bigger than the stroke) and used higher-specification materials than found on road bikes. Carburation was provided by twin Mikunis and the gases escaped via tuned pipes ending in a massive 'boom-box' muffler on the left of the bike.

This 'Mark 2' XR was to have a brief track success in the hands of Cal Rayborn in the Transatlantic series, but it was on the dirt tracks that the bike found its niche and main winning streak. The factory continued to make the major component parts for the bikes, at least as far as the engine was concerned, but allowed outside firms to make the rest of it, including the frame. In fact you couldn't buy a complete XR in the 1990s from Harley, it only sold the engine bits and advised buyers where to find the rest. A rider or tuner could then build the bike as they saw fit, as long as it met the minimum weight requirement of 315lb (143kg). Brake horsepower from most engines ended up around the 100 mark, more than enough to power the bike sideways!

Riders like Scott Parker, Jay Springsteen and the like, kept the XR750 alive on the loose surfaces, and to see one broadside at over 100mph (161km/h) is still a sight to behold. Ironically, the XR750, which had had such a troubled early life, ended up outliving both British and Japanese challenges to the traditional dirt track. American honour had been finally upheld.

ABOVE and OPPOSITE: Master and machine: Cal Rayborn on the Team Obsolete XR750 at Daytona.

SHOEI
vanson Leathers
TEAM OBSOLETE
Cal Rayborn
Domi Racer
KOSMAN
HARLEY-DAVIDSON
17
AVON TYRES
AirTECH
WORKS
TEAM OBSOLETE
MEGACYCLE

Drag Racing

Drag racing has long had an association with Harley-Davidson, right from the very first unofficial Saturday night races outside town, to the present day. This form of sport rapidly gained popularity in the US in the 1950s, taking over from hill climbs, although these still exist as a thriving branch of the the sport. But there was no official involvement until the late 1970s when, according to author David Wright, some of the proceeds from the sale of Aermacchi went into a factory drag-racing effort. Before that, Milwaukee seemed more keen on investing in outright speed records at the Bonneville Salt Flats with George Roeder piloting a streamlined Aermacchi Sprint 250 to 177.255mph (285.3km/h) in 1962 and Cal Rayborn managing over 245mph (394.3km/h) on an 89-ci (1460-cc) Sportster five years later.

Perhaps it was too late for Harley to dominate drag racing, but Marion Owens and a few others fought a rearguard action against multi-cylinder Japanese bikes in 1979 and 1980, Owens running a fearsome twin-engined machine. But even before then, Harleys were still prominent in US drag racing, with private riders campaigning successfully. Ron Fringer, for example, posted a 10.09-second ¼-mile time, with a 133.72mph (215.2km/h) terminal speed at Beech Bend Raceway in Kentucky, in 1971. Four years earlier, Les Payne had ducked below the 10-second barrier (9.71 seconds) with 152.38mph (245.2km/h) at Connecticut Raceway. Both used modified Sportster motors.

But even today, Harleys continue to do well in drag racing. This is due largely to the power characteristics of the big V-twin, which offers excellent torque from low revs, unlike the more 'peaky' four-cylinder opposition. UK specialist RMD of Reading, built a Harley-powered drag bike, and was delighted when it out-dragged a tuned Kawasaki ZZR1100. Although the Kawasaki posted a significantly higher terminal speed, and was quieter in the final stages of the ¼-mile, it was slower in those crucial first 200 yards (180m) from a standing start. It

LEFT: Only in America: skid racing involves removal of the entire front end and allowing the bike to rest on a skid plate. The rider places his weight as far back as possible, to minimize friction.

OPPOSITE: This stock-appearance bike and purpose-built dragster, both Harley-powered, make for an interesting comparison.

LEFT: A fearsome single-engined bike. Note the size of the exhaust pipes and carburettor intake!

OPPOSITE: Buells have been quickly accepted into drag racing. Wheelie bars, to prevent the bike flipping over backwards, are a mandatory fitment.

is this sort of performance that should keep Harleys well to the forefront of modern drag racing.

In fact, so popular have Harleys remained, that the American Drag Bike Association (www.americandragbike.com) runs its own drag series for Harley-engined bikes only. There are eight separate classes as follows:

MODIFIED – Largely standard Sportsters and Big Twins. Engines can be enlarged to 76ci/1245cc (Sportsters) or 84ci/1375cc (Big Twins) but no four-valve or twin carburettors are allowed. Engines must run on standard pump gas, and the lights and changing systems must be operational. The current class record is 9.355 seconds, 138.90mph (223.5km/h).

SUPER MODIFIED – Larger engines allowed, 76–120ci/1245–1965cc (Sportsters) or 84–120ci (Big Twins) but no four-valve or twin carburettor heads. Any aftermarket part is permitted as long as it is deemed safe by the ABDA. The transmission must be Harley-Davidson or a 'like same' replacement, gear ratios can be altered, but auto, rpm or computer-controlled changers are not allowed. An aftermarket frame is permitted, but only the sit-up type. The current class record is 8.522 seconds, 155.96mph (251km/h).

TOP GAS – Highly modified engines. These must be pushrod 45-degree V-twins, petrol-powered and naturally-aspirated, but

AHDRA
SEP 12

LEFT: Harley drag bikes are LOUD!

OPPOSITE: Underneath all the pipes and wires is a much-modified Evolution V-twin. The twin small-bore pipes are to operate the airshifter transmission, while the huge ones are for fuel.

any size induction and fuelling system is allowed. Any transmission is allowed, including rpm or computer shifting, and an aftermarket frame of any wheelbase. The current class record is 8.275 seconds, 157.53mph (253.5km/h).

PRO STOCK – Stock appearance with carburettors, XL, FX, or VR-based. Maximum 160ci (2620cc) on gas, but four-valve and twin carburettor heads are allowed. VRs must have factory cases and heads. Any transmission (including auto), 3–6-speeds, airshifters permitted. Racing slick tyres allowed, and maximum 70-in (178-cm) wheelbase. The current class record is 7.706 seconds, 166.13mph (267.3km/h).

PRO MODIFIED – Petrol or alcohol fuelling, naturally aspirated up to 151ci (2475cc), turbo or supercharged up to 122ci (2000cc). Any transmission (including auto), 68-in (173-cm) maximum wheelbase, any aftermarket frame (sit- or lay-down type) and V-rated or drag-racing tyres. The current class record is 7.915 seconds, 167.97mph (270.3km/h).

PRO DRAGSTER – Nitro-fuelled, carburettor, single engine up to 122ci (2000cc), pushrods, four-valve heads. No transmission, any clutch, any frame, 68-in (173-cm) maximum wheelbase. The current class record is 7.297 seconds, 174.75mph (281.2km/h).

LANSDALE
AMUSEMENT
COMPANY
Chassis By...

LEFT Drag bikes run over 1/8- or 1/4-mile, and although they run in pairs, the chief competitor is the clock. Start times are controlled by a 'Christmas Tree' of seven coloured lights (five amber, one red, one green), but the clock only starts and stops when the electronic timing beam is broken.

OPPOSITE: Smoking the rear tyre looks spectacular, but it's not for show, being the inevitable result of wheelspin.

PRO FUEL – Nitro-fuelled bikes designed specifically for drag racing. Maximum 151ci/2475cc (carburettors) or 132ci/2165cc (fuel injection), pushrods, four-valve heads. One- or two-speed transmission, any frame, maximum 72-in (183-cm) wheelbase, unlimited tyre size with single-speed transmission. The current class record is 6.921 seconds.

TOP FUEL – Nitro-fuelled with super or turbocharging. Single or double engines up to 175ci/2870cc combined. Must still use pushrods, but four-valve heads, any transmission or clutch allowed. 78-in (198-cm) maximum wheelbase and any tyres specified for drag racing. The current class record is 6.601 seconds, 213.52mph (343.6km/h).

SIMONS
CENTER
Buell
Pro Dragster
Top Fuel
S&S
AHDRA
101

DAVE'S PLACE
Indian
MOTORCYCLE
RED LINE OIL
RAM

OPPOSITE: The lay-down frame is used on purpose-built drag bikes, as the long wheelbase and forward weight distribution makes wheelies less likely.

RIGHT: Not strictly drag racing, this trike is engaged in a mild form of tractor pulling. It is also Harley-powered.

DOC'S
HARLEY-DAVIDSON
SHAWANO COUNTY, WIS.

OPPOSITE: A typical specification of a serious drag bike like this would read: 140-ci (2295-cc) motor, 3-inch (76.2-mm) belt drive, custom five-speed transmission, pneumatic shifting, chrome-moly tube frame, 10-in (254-mm) wide rear tyre on 15-in (381-mm) rim. The rider will be experiencing three times his own weight in accelerating 0–100 in less than 2 seconds.

RIGHT: This bike might look relatively stock, but underneath it's highly modified.

OPPOSITE: Notice that the whites of this rider's eyes are showing as he blasts off the line.

RIGHT: Blast-off! An ultimate nitro-fuelled drag bike.

AHDRA

OPPOSITE: Just to prove they're not all exotic and expensive, here is a stock-like 883 Sportster powering modestly away.

RIGHT: The very spirit of drag racing.

GUIDE TO MODEL NAMES

Model 0 – 1904, 24.74-ci (410-cc) F-head single
1 – 1905, 24.74-ci (410-cc) F-head single
2 – 1906, 26.8-ci (440-cc) F-head single
3 – 1907, 26.8-ci (440-cc) F-head single
4 – 1908, 26.8-ci (440-cc) F-head single
5/A/B/C – 1909, 30.5-ci (500-cc) F-head single
5D – 1909, 50-ci (820-cc) F-head V-twin, magneto
6/A/B/C/E – 1910, 30.5-ci (500-cc) F-head single
7/A/B/C – 1911, 30.5-ci (500-cc) F-head single
7D – 1911, 50-ci (820-cc) F-head V-twin
8/A/X – 1912, 30.5-ci (500-cc) F-head single
8D – 1912, 50-ci (820-cc) F-head V-twin
8E – 1912, 61-ci (1000-cc) F-head V-twin
9A/B – 1913, 35-ci (575-cc) F-head single
9E/F – 1913, 61-ci (1000-cc) F-head V-twin, chain drive
9G – 1913, 61-ci (1000-cc) F-head V-twin with front-mounted delivery box
10A/B/C – 1914, 35-ci (575-cc) F-head single
10E/F – 1914, 61-ci (1000-cc) F-head V-twin, chain drive
10G – 1914, 61-ci (1000-cc) F-head V-twin, Delivery Van, two-speed
11B/C – 1915, 35-ci (575-cc) F-head single
11E/F/H – 1915, 61-ci (1000-cc) F-head V-twin
11G – 1915, 61-ci (1000-cc) F-head V-twin Delivery Van
16B/C – 1916, 35-ci (575-cc) F-head single
16E/F – 1916, 61-ci (1000-cc) F-head V-twin
17B/C – 1917, 35-ci (575-cc) F-head single
17E/F – 61-ci (1000-cc) F-head V-twin

A – 1926–27, 21-ci (350-cc) side-valve single, magneto
A Topper scooter – 1960, 10-ci (165-cc) two-stroke single, automatic
AA – 1926–27 21-ci (350-cc) OHV single, magneto
AH Topper scooter – 1961–65, 10-ci (165-cc) two-stroke single, automatic
AU Topper scooter – 1960–64, 10-ci (165-cc) two-stroke single, automatic

B – 1918, 35-ci (375-cc) F-head single
B – 1926–34, 21-ci (350-cc) flat-head single, electrics
B – 1955–59, 7-ci (125-cc) two-stroke single
BA – 1926–29, 21-ci (350-cc) OHV single, electrics
BT – 1960–61, Super 10, 10-ci (165-cc) two-stroke single
BT Pacer – 1962–64, 11-ci (175-cc) two-stroke single
BTF Ranger – 1962, 10-ci (165-cc) two-stroke single
BTH Scat – 1962–65, 11-ci (175-cc) two-stroke single
BTH Bobcat – 1966, 11-ci (175-cc) two-stroke single
BTU Super 10 – 1960–61, 10-ci (165-cc) two-stroke single, restricted
BTU Pacer – 1962–64, 10-ci (165-cc) two-stroke single, restricted

C – 1918, 35-ci (575-cc) F-head single, magneto
C – 1929–34, 30.5-ci (500-cc) side-valve single, electrics
C Sprint – 1961–66, 15-ci (246-cc) OHV single
CB – 1933–34, 30.5-ci (500-cc) side-valve single
CD Commercial – 1921–22, 37-ci (606-cc) F-head single

D – 1929–31, 45-ci (750-cc) side-valve V-twin
DL – 1929–31, 45-ci (750-cc) flathead V-twin, high-compression
DLD – 1930–31, 45-ci (750-cc) side-valve V-twin, high-compression
DS – 1930–31, 45-ci (750-cc) side-valve V-twin, sidecar model

E – 1918, 61-ci (1000-cc) F-head V-twin
E – 1936–38 & 1944-50, 61-ci (1000-cc) OHV V-twin, medium-compression
EL – 1936–52, 61-ci (1000-cc) OHV V-twin, high-compression
ELF – 1952, 61-ci (1000-cc) OHV V-twin, foot-change
ELP – 1949, 61-ci (1000-cc) OHV V-twin, medium-compression, spring fork

ELS – 1951–52, 61-ci (1000-cc) OHV V-twin, medium-compression, sidecar gearing
EP – 1949, 61-ci (1000-cc) OHV V-twin, medium-compression, spring fork, sidecar gearing
ERS Sprint Scrambler – 1969–70, Sprint Scrambler, 15-ci (246-cc) single
ES – 1936–41, 1945–50, 61-ci (1000-cc) OHV V-twin, sidecar gearing

F – 1918–23, 61-ci (1000-cc) F-head V-twin, three-speed
F – 1942–49, 74-ci (1200-cc) OHV V-twin, medium-compression
F – 1950, 74-ci (1200-cc) OHV V-twin, medium-compression
FD – 1921–24, 74-ci (1200-cc) F-head V-twin, magneto
FDCA – 1924, 74-ci (1200-cc) F-head V-twin, alloy pistons, magneto
FDCB – 1925, 74-ci (1200-cc) F-head V-twin, iron-alloy pistons, magneto
FDCBS – 1925, 74-ci (1200-cc) F-head V-twin, iron-alloy pistons, magneto, sidecar gearing
FDS – 1921–24, 74-ci (1200-cc) F-head V-twin, magneto, sidecar gearing
FDSCA – 1924, 74-ci (1200-cc) F-head V-twin, alloy pistons, magneto, sidecar gearing
FE – 1924–25, 61-ci (1000-cc) F-head V-twin, alloy pistons, magneto
FES – 1924–25, 61-ci (1000-cc)F-head V-twin, alloy pistons, magneto, sidecar gearing
FL – 1941–64, 74-ci (1200-cc) OHV V-twin, high-compression
FL – 1973–75, 74-ci (1200-cc) OHV V-twin
FLB – 1965–69, 74-ci (1200-cc) OHV V-twin, hand-change
FLE – 1953–56, 74-ci (1200-cc) OHV V-twin, combination
FLEF – 1953–56, 74-ci (1200-cc) OHV V-twin, foot-change, combination
FLF – 1952–64, 74-ci (1200-cc) OHV V-twin, foot-change
FLFB – 1965–67, 74-ci (1200-cc) OHV V-twin, foot-change
FLH – 1955–64, 74-ci (1200-cc) OHV V-twin, high-compression
FLH – 1970–73, 74-ci (1200-cc) OHV V-twin, high-compression, hand-change
FLH – 1974–80, 74-ci (1200-cc) OHV V-twin
FLH-80 Electra Glide – 1978–84, 82-ci (1340-cc) OHV V-twin
FLH Electra Glide Heritage – 1981, 82-ci (1340-cc) OHV V-twin
FLHB Electra Glide – 1965–69, 74-ci (1200-cc) OHV V-twin, high-compression, hand-change
FLHC Electra Glide Classic – 1979–81, 82-ci (1340-cc) OHV V-twin
FLHF Hydra Glide – 1955–64, 74-ci (1200-cc) OHV V-twin, high-compression, foot-change
FLHF – 1970–73, 74-ci (1200-cc) OHV V-twin, high-compression, foot-change
FLHFB Electra Glide – 1965–69, 74-ci (1200-cc) OHV V-twin, high-compression, foot-change
FLHP Road King Police – 1998, 82-ci (1340-cc) Evo V-twin
FLHPI Road King Police – 1998, 82-ci (1340-cc) Evo V-twin, e.f.i.
FLHR Electra Glide Road King – 1994–98, 82-ci (1340-cc) Evo V-twin
FLHR Road King – 1999–2001, 88-ci (1450-cc) Twin Cam V-twin
FLHRCI Road King Classic – 1998, 82-ci (1340-cc) Evo V-twin e.f.i.
FLHRCI Road King Classic – 1999–2001, 88-ci (1450-cc) Twin Cam V-twin e.f.i.
FLHRI Electra Glide Road King – 1996–98, 82-ci (1340-cc) Evo V-twin e.f.i.
FLHS Electra Glide – 1977, 74-ci (1200-cc) OHV V-twin
FLHS Electra Glide – 1980–81, 82-ci (1340-cc) OHV V-twin
FLHS Electra Glide Sport – 1983–93, 82-ci (1340-cc) OHV V-twin
FLHT Electra Glide – 1983, 82-ci (1340-cc) OHV V-twin
FLHT Electra Glide Standard – 1995–98, 82-ci (1340-cc) Evo V-twin
FLHT Electra Glide Standard – 1999, 88-ci (1450-cc) Twin Cam V-twin
FLHT Electra Glide Standard – 2001 on, 88-ci (1450-cc) Twin Cam V-twin
FLHTC Electra Glide Classic – 1983–97, 82-ci (1340-cc) OHV V-twin
FLHTC Electra Glide Classic – 1998, 82-ci (1340-cc) Evo V-twin
FLHTCI Electra Glide Classic – 1996–

98, 82-ci (1340-cc) Evo V-twin e.f.i.
FLHTCU Electra Glide Classic Ultra – 1989–97, 82-ci (1340-cc) Evo V-twin
FLHTCUI Electra Glide Classic Ultra – 1995–98, 82-ci (1340-cc) Evo V-twin e.f.i.
FLHTCUI Electra Glide Classic Ultra – 2000 on, 88-ci (1450-cc) Twin Cam V-twin e.f.i.
FLHTPI Electra Glide Police – 1998, 82-ci (1340-cc) Evo V-twin e.f.i.
FLHX Electra Glide – 1984, 82-ci (1340-cc) OHV V-twin
FLP – 1949, 74-ci (1200-cc) OHV V-twin, high-compression, spring fork
FLP – 1970–72, 74-ci (1200-cc) OHV V-twin, hand-change
FLPF– 1970–72, 74-ci (1200-cc) OHV V-twin, foot-change
FLS Hydra-Glide – 1951–52, 74-ci (1200-cc) OHV V-twin, medium-compression, sidecar model
FLST Heritage Softail – 1986–90 82-ci (1340-cc) V-twin
FLSTC Heritage Softail Special – 1987–98, 82-ci (1340-cc) Evo V-twin
FLSTC Heritage Softail Classic – 2000 on, 88-ci (1450-cc) Twin Cam 88B V-twin
FLSTCI Heritage Softail Classic – 2001 on, 88-ci (1450-cc) Twin Cam 88B V-twin e.f.i.
FLSTF Fat Boy – 1990–99, 82-ci (1340-cc) Evo V-twin
FLSTF Fat Boy – 2000 on, 88-ci (1450-cc) Twin Cam 88B V-twin
FLSTFI Fat Boy – 2001 on, 88-ci (1450-cc) Twin Cam 88B V-twin e.f.i.
FLSTN Heritage Softail Nostalgia – 1993–95, 82-ci (1340-cc) Evo V-twin
FLSTN Heritage Softail Special – 1996, 82-ci (1340-cc) Evo V-twin
FLSTS Heritage Springer – 1997–98, 82-ci (1340-cc) Evo V-twin
FLSTS Heritage Springer – 2001 on, 88-ci (1450-cc) Twin Cam 88B V-twin
FLT Tour Glide – 1980–82, 82-ci (1340-cc) OHV V-twin
FLTC Tour Glide Classic– 1981–94, 82-ci (1340-cc) OHV V-twin
FLTCU Tour Glide Classic Ultra – 1989–95, 82-ci (1340-cc) Evo V-twin
FLTCUI Tour Glide Classic Ultra – 1996–97, 82-ci (1340-cc) Evo V-twin e.f.i.
FLTR Road Glide – 1998, 82-ci (1340-cc) Evo V-twin
FLTRI Road Glide – 1998, 82-ci (1340-cc) Evo V-twin e.f.i.
FLTRI Road Glide – 1999–2001, 88-ci (1450-cc) Twin Cam V-twin e.f.i.
FP – 1949, 74-ci (1200-cc) OHV V-twin, medium-compression, spring fork, sidecar gearing
FS – 1919–23, 61-ci (1000-cc) F-head V-twin sidecar motor
FS – 1941, 1945–50, 74-ci (1200-cc) OHV V-twin, sidecar gearing
FUS – 1918, 61-ci (1000-cc) F-head V-twin, Presto-lite
FX Super Glide – 1971–78, 74-ci (1200-cc) OHV V-twin
FXB Sturgis – 1980–82, 82-ci (1340-cc) OHV V-twin
FXD Dyna Super Glide – 1995–98, 82-ci (1340-cc) Evo V-twin
FXD Dyna Super Glide – 1999–2001, 88-ci (1450-cc) Twin Cam V-twin
FXDB Dyna Glide Sturgis – 1991, 82-ci (1340-cc) Evo V-twin
FXDB Dyna Glide Daytona – 1992, 82-ci (1340-cc) Evo V-twin
FXDC Dyna Glide Custom – 1992, 82-ci (1340-cc) Evo V-twin
FXDL Dyna Low Rider – 1994–98, 82-ci (1340-cc) Evo V-twin
FXDL Dyna Low Rider – 1999–2001, 88-ci (1450-cc) Twin Cam V-twin
FXDG Disc Glide – 1983, 82-ci (1340-cc) OHV V-twin
FXDWG Dyna Wide Glide – 1993–98, 82-ci (1340-cc) Evo V-twin
FXDWG Dyna Wide Glide – 1999–2001, 88-ci (1450-cc) Twin Cam V-twin
FXDX Dyna Super Glide Sport – 1999–2001, 88-ci (1450-cc) Twin Cam V-twin
FXDXT Dyna Super Glide T-Sport – 2001 on, 88-ci (1450-cc) Twin Cam V-twin
FXWG Wide Glide – 1980–86, 82-ci (1340-cc) OHV V-twin
FXE Super Glide – 1974–80, 74-ci (1200-cc) OHV V-twin, electric start
FXE Super Glide – 1981–84, 82-ci (1340-cc) OHV V-twin, electric start
FXEF-1200 Fat Bob – 1979–80, 74-ci (1200-cc) OHV V-twin
FXEF-80 Fat Bob – 1979–81, 82-ci (1340-cc) OHV V-twin
FXEF Fat Bob – 1985, 82-ci (1340-cc) V-twin
FXLR Low Rider Custom – 1987–94, 82-ci (1340-cc) Evo V-twin
FXR Super Glide II – 1982–83, 82-ci (1340-cc) OHV V-twin
FXR Super Glide – 1986–94, 82-ci (1340-cc) Evo V-twin
FXRC Low Glide Custom – 1985, 82-ci (1340-cc) V-twin
FXRD Sport Glide Grand Touring – 1986, 82-ci (1340-cc) V-twin
FXRDG Disc Glide – 1984, 82-ci (1340-cc) OHV V-twin
FXRP Police – 1984, 82-ci (1340-cc) V-twin
FXRS Super Glide II – 1982, 82-ci (1340-cc) OHV V-twin
FXRS Low Glide – 1984–88, 82-ci (1340-cc) Evolution V-twin
FXRS Low Rider – 1990–91, 82-ci (1340-cc) Evo V-twin

FXRS-CONV Low Rider Convertible – 1989–98, 82-ci (1340-cc) Evo V-twin
FXRS-SP Low Rider Sport – 1987–93, 82-ci (1340-cc) Evo V-twin
FXRT Sport Glide – 1983–92, 82-ci (1340-cc) OHV V-twin
FXS Low Rider – 1977–80, 74-ci (1200-cc) OHV V-twin
FXS-80 Low Rider – 1979–82, 82-ci (1340-cc) OHV V-twin
FXSB Low Rider – 1983–85, 82-ci (1340-cc) OHV V-twin
FXST Softail – 1984–90, 82-ci (1340-cc) Evo V-twin
FXST Softail Standard – 1999–2001, 88-ci (1450-cc) Twin Cam V-twin
FXSTB Night Train – 1998–99, 82-ci (1340-cc) Evo V-twin
FXSTB Night Train – 2000 on, 88-ci (1450-cc) Twin Cam 88B V-twin
FXSTC Softail Custom – 1986–98, 82-ci (1340-cc) Evo V-twin
FXSTD Softail Deuce – 2000 on, 88-ci (1450-cc) Twin Cam V-twin
FXSTDI Softail Deuce – 2001 on, 88-ci (1450-cc) Twin Cam 88B V-twin, e.f.i.
FXSTS Springer Softail – 1988–99 82-ci (1340-cc) Evo V-twin
FXSTS Springer Softail – 2001 on, 88-ci (1450-cc) Twin Cam 88B
FXSTSB Bad Boy – 1995–97, 82-ci (1340-cc) Evo V-twin

G Servicar – 1932–63, 45-ci (750-cc) side-valve V-twin
GD Servicar – 1932–41, 45-ci (750-cc) side-valve V-twin, large compartment
GE Servicar – 1933–37, 45-ci (750-cc) side-valve V-twin, air tank
GE Servicar – 1964–73, 45-ci (750-cc) side-valve V-twin, electric start

H Sprint H - 1962-68, 246cc OHV single, high compression

J – 1915–29, 61-ci (1000-cc) F-head V-twin, three-speed, electrics
JD – 1921–29, 74-ci (1200-cc) F-head V-twin, electrics
JDCA – 1924, 74-ci (1200-cc) F-head V-twin, alloy pistons, electrics
JDCB – 1925, 74-ci (1200-cc) F-head V-twin, iron-alloy pistons, electrics
JDCBS – 1925, 74-ci (1200-cc) F-head V-twin, iron-alloy pistons, electrics, sidecar gearing
JDH Two Cam – 1928–29, 74-ci (1200-cc) F-head V-twin, Dow-metal pistons, electrics
JDS – 1921–29, 74-ci (1200-cc) F-head V-twin, three-speed, electrics, sidecar gearing
JDSCA – 1924, 74-ci (1200-cc) F-head V-twin, alloy pistons, electrics, sidecar gearing
JDX– 1928, 74-ci (1200-cc) F-head V-twin, electrics
JDXL – 1928, 74-ci (1200-cc) F-head V-twin, Dow-metal pistons, electrics
JE – 1924–25, 61-ci (1000-cc) F-head V-twin, alloy pistons, electrics
JES – 1924–25, 61-ci (1000-cc) F-head V-twin, alloy pistons, electrics, sidecar gearing
JH Two Cam – 1928–29, 61-ci (1000-cc) F-head V-twin, Dow-metal pistons, electrics
JS – 1919–29, 61-ci (1000-cc) F-head V-twin sidecar motor, electrics
JX – 1928, 61-ci (1000-cc) F-head V-twin, three-speed, electrics
JXL – 1928, 61-ci (1000-cc) F-head V-twin, Dow-metal pistons, electrics

K – 1952–53, 45-ci (750-cc) side-valve V-twin
KH – 1954–56, 55-ci (900-cc) side-valve V-twin
KHK –1955–56, 55-ci (900-cc) side-valve V-twin, tuned
KK – 1953, 45-ci (750-cc) side-valve V-twin, tuned
M50 –1965–67, 3-ci (49-cc) two-stroke single
M65 – 1967–69, 4-ci (65-cc) two-stroke single
MC65 Shortster – 1972, 4-ci (65-cc) two-stroke single
M125 Rapido – 1968–72, 7.5-ci (124-cc), two-stroke single
MS50 – 1966–68, 3-ci (49-cc) two-stroke single
MS65 – 1967–72, 4-ci (65-cc) two-stroke single
MSR Baja – 1970–72, 6-ci (98-cc) two-stroke single

R – 1932–36, 45-ci (750-cc) side-valve V-twin, low-compression
RC125 – 1974, 7.5-ci (124-cc) two-stroke single
RL – 1932–36, 45-ci (750-cc) side-valve V-twin
RLD – 1932–36, 45-ci (750-cc) side-valve V-twin, high-compression
RS – 1932–36, 45-ci (750-cc) side-valve V-twin, sidecar model

S – 1926–27, 21-ci (350-cc) OHV single racer, magneto
S – 1948–52, 7.5ci (125-cc) two-stroke single
SR100 – 1973–74, 6-ci (98-cc) two-stroke single
SS Sprint SS – 1967–71, 15-ci (246-cc) OHV single
SS Sprint SS – 1973–74, 21-ci (346-cc) OHV single
SS125 – 1976–77, 7-ci (124-cc) two-stroke single
SS175 – 1976–77, 10.5-ci (174-cc)

OPPOSITE: Riding into the millennium.

two-stroke single
SS250 – 1975–77, 14.8-ci (242-cc) two-stroke single
ST – 1953–59, 10-ci (165-cc) two-stroke single
STU – 1954–59, 10-ci (165-cc) two-stroke single, restricted
SX Sprint SX – 1971–74, 21-ci (346-cc) OHV single
SX 125 – 1974–75, 7.5-ci (124-cc) two-stroke single
SX 175 – 1974–75, 10.5-ci (174-cc) two-stroke single
SX 250 – 1975–78, 14.8-ci (242-cc) two-stroke single
SXT 125 – 1976–77, 7.5-ci (124-cc) two-stroke single

TX 125 – 1973, 7.5-ci (124-cc) two-stroke single

U – 1937–48, 74-ci (1200-cc) side-valve V-twin, low-compression
UC – 1945, 74-ci (1200-cc) side-valve V-twin, medium-compression
UH – 1937–41, 80-ci (1300-cc) side-valve V-twin, medium-compression
UHS – 1937–41, 80-ci (1300-cc) side-valve V-twin, sidecar gearing
UL – 1937–48, 74-ci (1200-cc) side-valve V-twin, high-compression
ULH – 1937–41, 80-ci (1300-cc) side-valve V-twin, high compression
US – 1937–41, 1945–48, 74-ci (1200-cc) side-valve V-twin, sidecar gearing

V – 1930–33, 74-ci (1200-cc) side-valve V-twin, medium-compression
VC – 1930–33, 74-ci (1200-cc) side-valve V-twin, commercial model
VD – 1934–36, 74-ci (1200-cc) side-valve V-twin, low-compression
VDDS – 1935, 80-ci (1300-cc) side-valve V-twin, sidecar model
VDS – 1934–36, 74-ci (1200-cc) side-valve V-twin, low-compression, sidecar gearing
VHS – 1936, 80-ci (1300-cc) side-valve V-twin, sidecar gearing
VL – 1930–33, 74-ci (1200-cc) side-valve V-twin, high-compression
VLD – 1933–35, 74-ci (1200-cc) side-valve V-twin, high compression
VLDD – 1935, 80-ci (1300-cc) side-valve V-twin, high compression
VLH – 1936, 80-ci (1300-cc) side-valve V-twin, high compression
VS – 1930–33, 74-ci (1200-cc) side-valve V-twin, sidecar model

W – 1937, 45-ci (750-cc) side-valve V-twin, low-compression
WF – 1919–23, 35-ci (575-cc) ho twin, three-speed, magneto
WJ – 1919–23, 35-ci (575-cc) ho twin, three-speed, electrics
WL – 1937–42, 1945–51, 45-ci (750-cc) side-valve V-twin, high-compression
WLA – 1941–45, 45-ci (750-cc) side-valve V-twin, US Army
WLC – 1941–44, 45-ci (750-cc) side-valve V-twin, Canadian Army
WLD – 1937–42, 45-ci (750-cc) side-valve V-twin, extra-high compression
WLR – 1945, 45-ci (750-cc) side-valve V-twin, Russian Army
WS – 1937, 45-ci (750-cc) side-valve V-twin, sidecar gearing

X90 – 1973–75, 5.5-ci (90-cc) two-stroke single
XA – 1942–43, 45-ci (750-cc) flathead ho twin, US Army
XL Sportster – 1957–59, 54-ci (883-cc) OHV V-twin, medium-compression
XL53C Sportster Custom 53 – 1998–2001, 54-ci (883-cc) Evo V-twin
XL883 Sportster - 2000 on, 54-ci (883-cc) Evo V-twin
XL883 Sportster Hugger – 2000 on, 54-ci (883-cc) Evo V-twin
XL1200C Sportster Custom – 1996–2001, 73-ci (1203-cc) Evo V-twin
XL1200S Sportster Sport –1996–2001, 73-ci (1203-cc) Evo V-twin
XLC Sportster – 1958–59, 54-ci (883-cc) OHV V-twin, high-compression
XLH Sportster – 1958–71, 54-ci (883-cc) OHV V-twin, high-compression
XLH Sportster – 1972–85, 61-ci (997-cc) OHV V-twin
XLH Sportster – 1986–98, 54-ci (883-cc) Evo V-twin
XLH Sportster – 1988–99, 73-ci (1203-cc) Evo V-twin
XLH Sportster Hugger – 1988–99, 54-ci (883-cc) Evo V-twin
XLH Sportster Deluxe – 1988–95, 54-ci (883-cc) Evo V-twin
XLH1100 Sportster – 1986–87, 67-ci (1100-cc) V-twin
XLCH Sportster – 1958–71, 54-ci (883-cc) OHV V-twin, high-compression
XLCH Sportster – 1972–79, 997cc OHV V-twin
XLCR Cafe Racer -1977-78, 61-ci (997-cc) OHV V-twin
XLS Sportster/Roadster – 1979–85, 61-ci (997-cc) OHV V-twin
XLT Sportster – 1977–78, 61-ci (997-cc) OHV V-twin
XLX Sportster – 1983–85, 61-ci (997-cc) OHV V-twin
XR1000 Sportster – 1983–84, 61-ci (1000-cc) OHV V-twin

Z90 – 1973–75, 5.5-ci (90-cc) two-stroke single

PARKING
TEXACO

INTO THE MILLENNIUM

ABOVE and OPPOSITE: For 2000, a genuine factory custom limited edition by Black Bear of the UK.

BLACK BEAR
HARLEY-DAVIDSON
HARLEY-DAVIDSON

2000 Range

RIGHT: A Buell SI Lightning at Daytona.

OPPOSITE: A Buell SI Lightning.

VANCE & HINES

BELOW: A Sportster XL 883.

OPPOSITE: A Sportster XL1200S

Sportster
Harley-Davidson
1200
BELL

RIGHT and OPPOSITE: Softail Standard.

Softail with custom accessories.

Softail with custom add-ons.

RIGHT and BELOW: Dyna Super Glide.

OPPOSITE: A Dyna Wide Glide at the Daytona 2000.

PIER
WELCOME TO DAYTONA BEACH

ABOVE and OPPOSITE: Dyna Super Glide Sport.

HARLEY-DAVIDSON

RIGHT: Road King.

OPPOSITE: Road King Classic.

HARLEY-DAVIDSON

RIGHT and OPPOSITE: Ultra Classic Electra Glide.

SUPER UNLEADED
Coors LIGHT
WELCOME
BURGERS

HARLEY-DAVIDSON
Touring

OPPOSITE: Ultra Classic Electra Glide.

RIGHT: Heritage Softail.

Three Road Glides at Sturgis.

ABOVE: Road Glide.

N ROCHFORD
S ROCHFORD
US 38
HARLEY-DAVIDSON

OPPOSITE and BELOW: Electra Glide Classics.

Screamin' Eagle Road Glide.

Limited edition Screamin' Eagle Road Glide.

Y2K

OPPOSITE and THIS PAGE
Y2K Khaos Millennium bike, customized by Cory Ness.

2001 Range

Sportster Range
Sportster 883 (54ci/883cc)
Sportster 883 Hugger (54ci/883cc)
Sportster 53C Custom (54ci/883cc)
Sportster 1200 Custom (74ci/1199cc)
Sportster 1200 Sport (74ci/1199cc)

Dyna Glide Range
Super Glide (88ci/1449cc)
Low Rider (88ci/1449cc)
Wide Glide (88ci/1449cc)
Super Glide Sport (88ci/1449cc)
Super Glide T-Sport (88ci/1449cc)

Softail Range
Softail Standard (88ci/1449cc)
Night Train (88ci/1449cc)
Fat Boy (88ci/1449cc)
Fat Boy EFI (88ci/1449cc)
Heritage Classic (88ci/1449cc)
Heritage Classic EFI (88ci/1449cc)
Softail Deuce (88ci/1449cc)
Softail Deuce EFI (88ci/1449cc)

Touring Range
Road King (88ci/1449cc)
Road King Classic (88ci/1449cc)
Ultra Classic Electra Glide (88ci/1449cc)

RIGHT and OPPOSITE: XL Sportster 883.

BELOW: Dyna Super Glide Sport.
OPPOSITE: Dyna Super Glide T-Sport.

HARLEY
DAVIDSON

Harley-Davidson

OPPOSITE: Softail Deuce.

RIGHT: Fat Boy Softail.

RIGHT: Road King Classic.

OPPOSITE: Ultra Classic Electra Glide.

2002

Harley-Davidson V-Rod: An American Revolution

Every 20 years or so, Harley-Davidson comes up with a new engine, or at least a new development of what has gone before. Consequently, the overhead-valve Knucklehead of 1936 owed much to previous side-valve Harley V-twins; the Panhead (1947) and Shovelhead (1996) which succeeded it were all developments of the same thing. Even the Evolution of 1983 – a great leap forward for Harley – was based on the same old bottom end, albeit with lots of new parts. But in the summer of 2001 came something completely new – the V-Rod.

At the dawn of the 21st century, Harley-Davidson was selling more bikes and making bigger profits than ever before. But it couldn't afford to rest on its laurels. Quite apart from ever more sophisticated (and authentic-looking) Japanese cruisers, there was homegrown opposition from the Victory. The V-Rod was the answer – the first all-new Harley-Davidson for over half a century. All previous Harley road twins had been based on the same air-cooled 45-degree format, but the V-Rod was completely different. Not for nothing was the new engine named 'Revolution'.

The 60-degree angle between the cylinders was a huge break with tradition, as was the use of water-cooling and four valves per cylinder. It is tempting to add fuel injection and a balance shaft to that list (the V-Rod has both), but Harley-Davidson had already applied these to the faithful Evolution V-twin. By Harley standards, at 1130cc the new engine was surprisingly small compared to the 1450-cc (88ci) Twin Cam. But it was more powerful than any road engine Harley-Davidson had ever put its name to, with 115bhp at 8,000rpm. This was the sort of power needed to counter the latest generation of sports cruisers, and the V-Rod delivered.

But the new engine wasn't quite as fresh as it seemed, being a much changed, sanitized version of the VR1000 race motor. As a racer, that was a flop (it secured a single place and no wins in seven years of trying), but as the basis for a new generation of road bikes it looks more promising.

Harley-Davidson recruited sports car maker Porsche to transform the VR1000 engine into a 21st-century V-twin, and build the first prototypes. Harley finalized the R&D on the engine, as well as designing the chassis that would house it. However, the new engine is a product of a joint venture between Porsche and Harley, though built in Harley-Davidson's own Kansas City factory.

In comparison with this revolutionary engine (by Harley standards), the chassis was decidedly conservative: tubular steel (not alloy) frame, with short-travel rear shocks adjustable only for pre-load; the front forks weren't adjustable at all, and the concerning clearance was a mere 32 degrees; like most Harleys, even this one would soon ground out if the rider got too enthusiastic on twisty roads! However, most people agree that there at last is a half-modern-looking Harley, power-packed and muscular, all bright silver and polished alloy. And who led the styling team? Wille G. Davidson. Respected motorcycle journalist Alan Cathcart summed it all up after riding the V-Rod in August 2001: 'This is one of the most important motorcycles of the past decade – Harley-Davidson has reinvented itself.'

SPECIFICATIONS
Harley-Davidson VRSC V-Rod

Engine	Liquid-cooled 60-degree V-twin, DOHC 4 valves/cylinder
Bore x stroke	100 x 72mm
Capacity	1130cc (69ci)
Compression ratio	11.3:1
Power	115bhp @ 8,000rpm
Torque	65lb ft @ 6,300
Gearbox	Five-speed
Frame Type	Tubular steel
Rake/Trail	34 degrees/99mm
Wheelbase	1713mm
Suspension	F: 49mm forks, non-adj, 100mm travel R: Twin shocks, pre-load, adj, 60mm travel
Brakes	F: Twin discs, 292mm R: Disc, 292mm
Tyres	F: 120/70 19 R: 180/55 18
Seat Height	686mm
Fuel Capacity	15 litres (3.3)
Weight (dry)	270kg (595lb)

HARLEY-DAVIDSON
V-ROD

2002-2010

Sportster Range for 2009
Sportster Low 883 (54ci/883cc)
Sportster Custom 883 (54ci/883cc)
Sportster Iron 883 (54ci/883cc)
Sportster Low 1200 (74ci/1199cc)
Sportster Nightster
Sportster Low 1200 (74ci/1199cc)
Sportster Custom 1200 (74ci/1199cc)
Sportster XR 1200 (74ci/1199cc)

Dyna Glide Range for 2009
Super Glide (96ci/1584cc)
Super Glide Custom (96ci/1584cc)
Street Bob (96ci/1584cc)

Low Rider (96ci/1584cc)
Fat Bob (96ci/1584cc)

Softail Range for 2009
Softail (96ci/1584cc)
Night Train (96ci/1584cc)
Fat Boy (96ci/1584cc)
Softail Delux (96ci/1584cc)
Heritage Softail Classic (96ci/1584cc)
Heritage Classic EFI (96ci/1584cc)
Rocker (96ci/1584cc)
Cross Bones (96ci/1584cc)

VRSC Range for 2009
V-Rod (76ci/1584cc)
Night Rod Special (76ci/1584cc)
V Rod Muscle (76ci/1584cc)

Touring Range
Road King (96ci/1584cc)
Road King Classic (96ci/1584cc)
Street Glide (96ci/1584cc)
Road Glide (96ci/1584cc)
Electra Glide Standard (96ci/1584cc)
Ultra Classic Electra Glide (96ci/1584cc)
Tri Glide Classic (96ci/1584cc)

OPPOSITE LEFT: A Sportster Nightster. Picture courtesy of Flickr/Carbob49.

OPPOSITE RIGHT: Two Heritage Softail Delux. Picture courtesy of Flickr/OmniNate.

ABOVE: A 2007 Custom Sportster. Picture courtesy of Flickr/Kwalk628.

ABOVE LEFT: Road King part of thr Touring range. Picture courtesy of Flickr/Worldwidemotorcycles.

LEFT: A V-Rod from the VRSC range. Picture courtesy of Flickr/Elite PhotoArt.

Page numbers in italics refer to illustrations

A book as comprehensive as this is never possible without the help of a great many people. In particular, the authors would like to thank Garry Stuart for his excellent photographs, also Dr Martin Rosenblum for permission to reproduce his Harley-Davidson poems.

All the photographs in this book are courtesy of Gary Stuart, with the exception of those on pages 6, 7, 9, 14, 24 right, 191, 192, and 430–439, which are courtesy of Harley-Davidson UK: page 70 above left which is by John Carroll: pages 300 right, 301, 306, 311, 313–319 which are by Ian Kerr.

Garry Stuart wishes to thank the following people and organizations for their help and co-operation:
Glenn Bator, The Otis Chandler Museum, Oxnard, California; Lawayne Mathies, Mystery Design, Dallas, Texas; Dale and Martha of Greeley, Colorado; Martin Henderson, Boothill Motorcycles, England; Lonnie Isam Competition Motorcycles, Pearland, Texas; Mr Cole, Cole Motorcycle Museum, Sturgis, South Dakota;
Harley-Davidson UK Press PR, Brackley, UK; John and Helen Stuart, Holyhead, UK; Advantage Photo Services, Swindon, UK; and Harley-Davidson riders everywhere.

In memory of Dr 'Maz' Harris, HAMC, Kent, UK.